Bibliografische Information der Deutschen Nationalbibliothek

Die Deutsche Nationalbibliothek verzeichnet diese Publikation in der
Deutschen Nationalbibliografie; detaillierte bibliografische Daten sind
im Internet über http://dnb.d-nb.de abrufbar.

ISBN 978-3-8325-2617-7

Logos Verlag Berlin GmbH
Comeniushof, Gubener Str. 47,
10243 Berlin
Tel.: +49 (0)30 42 85 10 90
Fax: +49 (0)30 42 85 10 92
INTERNET: http://www.logos-verlag.de

Development of predictive tools for the characterisation of milling behaviour of pharmaceutical powders

Entwicklung von Methoden zur Vorhersage der Zerkleinerungseigenschaften pharmazeutischer Pulver

Der Technischen Fakultät der
Universität Erlangen-Nürnberg
zur Erlangung des Grades

DOKTOR-INGENIEUR

vorgelegt von

Matthias Meier

Erlangen - 2010

Als Dissertation genehmigt von

der Technischen Fakultät der

Universität Erlangen-Nürnberg

Tag der Einreichung: 09.10.2009

Tag der Promotion: 14.07.2010

Dekan: Prof. Dr.-Ing. habil. R. German

Berichterstatter: Prof. Dr.-Ing. W. Peukert

   Prof. Dr. T. Pöschel

Meinen Eltern

Vorwort

Die vorliegende Arbeit entstand in den Jahren 2003 bis 2008 während meiner Tätigkeit als wissenschaftlicher Mitarbeiter am Lehrstuhl für Feststoff- und Grenzflächenverfahrenstechnik an der Friedrich-Alexander-Universität Erlangen-Nürnberg in Kooperation mit der Novartis Pharma AG in Basel.

Mein besonderer Dank gilt Herrn Prof. Dr.-Ing. Wolfgang Peukert für die Möglichkeit der Durchführung dieser Promotion und für sein stetes Interesse an meiner Arbeit. Seine wertvollen Anregungen haben wesentlich zum Gelingen dieser Arbeit beigetragen. Herrn Prof. Dr. Thorsten Pöschel danke ich für die bereitwillige Übernahme des Koreferats. Dem fachfremden Gutachter Herrn Prof. Dr. Mathias Göken sei ebenso gedankt wie Herrn Prof. Dr. Axel König für die Übernahme des Prüfungsvorsitzes.

Mein freundschaftlicher Dank gilt den Kooperationspartnern Herrn Dr. Wolfgang Wirth, Herrn Dr. Edgar John und Herrn Dierk Wieckhusen von Novartis für die Betreuung meiner Arbeit von industrieller Seite, für die stets offenen Diskussionen und für die immer vorhandene Unterstützung während meiner Promotionszeit. Für die bereitwillige Unterstützung meiner Experimente in Basel geht ein besonderer Dank an die Herren Anton Baumberger, Eric Wirth, Bernard Peter, Hans-Peter Heim und Andre Knoll von der Mahlstation, Peter Kamber und Alain Merlo aus der Werkstatt, Martin Frei, Sven Mohler, Armin Katzenstein, Markus Mathys, Kurt Paulus, Dirk Märtin, Sebastian Kärger, Frank Stowasser und Phillippe Piechon aus der Partikelcharakterisierung und Thorsten Hummelsiep von der Kristallisation.

Bei allen Freunden, Kollegen und Mitarbeitern am Lehrstuhl möchte ich mich für die kollegiale und freundschaftliche Atmosphäre bedanken. Neben dem angenehmen Arbeitsklima werde ich die Zeit am Lehrstuhl auch durch die zahlreichen gemeinsamen Aktivitäten außerhalb des Universitätsalltags in guter Erinnerung behalten. Insbesondere den Herren Heinrich Drost und Peter Rollig gebührt Dank für die schnelle und zuverlässige Umsetzung zahlreicher größerer und kleinerer Konstruktionen und Verbesserungen. Frau Sieglinde Winter sei für Ihre Unterstützung in organisatorischen Fragen gedankt. Für ihren Einsatz und die gewissenhafte Unterstützung bei der Bearbeitung von Teilprojekten dieser Arbeit bedanke ich mich ganz herzlich bei meinen Diplomarbeitern, Studienarbeitern und studentischen Hilfskräften: Sonja Simon, Amol Dhumal, Sandra Wilhelm, Jin Geng, Melina Machado, Gabriela Guevara, Thomas Friemert, Andreas Frey und Katharina Stark.

Abschließend möchte ich ganz herzlich meinen Eltern danken, die mir die Ausbildung meiner Wahl ermöglicht und mich auf meinem Weg stets unterstützt haben.

Oppau, im August 2010   Matthias Meier

Abstract

Comminution is an important process step in the formulation of pharmaceutical substances, as the correct size of the drug particles plays a decisive role in their application. In the pharmaceutical industry, new molecular entities with yet unknown properties are generated permanently, thus also the milling process has often to be adapted. Usually, pilot-scale milling trials are done with large amounts of substance. This means a costly procedure, and these trials can be done only relatively late in the development process, when those amounts are available.

The aim of this work is to develop methods to characterise the comminution behaviour of new pharmaceutical substances quicker, and using less material. Two different aspects were considered: on the one hand, the intrinsic breakage behaviour of the substance, on the other hand its transport properties. These are governed by particle-particle and particle-wall interactions and can also influence the comminution process. For the characterisation of breakage behaviour, single particle comminution experiments were done in a new apparatus. From these measurements, parameters are obtained that describe the breakage probability of the substances. Further on, it could be shown that these parameters also determine the breakage function, i. e. the size distribution after comminution, and thus enable a complete description of the breakage behaviour. Mechanical properties of the substances were determined by nanoindentation and compared to the breakage parameters. Thus it could be shown that there is a direct correlation between the brittleness (defined as the ratio of hardness to fracture toughness) of a substance and its comminution behaviour.

For the characterisation of transport properties and adhesive and caking behaviour, shear tests were done, and surface properties were studied by atomic force microscopy (AFM) and inverse gas chromatography. In these experiments, no material specific properties could be identified that were responsible for enhanced particle adhesion. Rather, geometric factors as, for example, smooth or rough surfaces, seem to dominate adhesion. Generally, it is difficult to transfer results from AFM to macroscopic processes such as shear tests or milling experiments, because the loads in these cases are very different.

Zusammenfassung

Die Zerkleinerung ist ein wichtiger Prozeßschritt bei der Formulierung pharmazeutischer Stoffe. Die Einstellung der richtigen Partikelgröße ist von großer Bedeutung für die Wirksamkeit eines Medikaments. Da in der pharmazeutischen Industrie ständig neue Stoffe mit noch unbekannten Eigenschaften hergestellt werden, ist auch die Mahlung stets an neue Stoffe anzupassen. Hierzu werden üblicherweise Mahlversuche im Technikums-maßstab mit Stoffmengen bis zu mehreren Kilogramm gefahren. Dies bedeutet erhebliche Kosten; zudem können solche Mahlversuche erst in einem fortgeschrittenen Stadium der Entwicklung durchgeführt werden, wenn entsprechende Mengen zur Verfügung stehen.

Ziel dieser Arbeit ist es, Methoden zu finden, um das Zerkleinerungsverhalten neuer pharmazeutischer Substanzen schneller und mit weniger Materialeinsatz charakterisieren zu können. Hierfür wurden zwei unterschiedliche Aspekte betrachtet: einmal das tatsächliche Bruchverhalten des Stoffes, zum anderen die Transporteigenschaften, welche durch interpartikuläre Wechselwirkungen bestimmt werden, und die ebenfalls den Zerkleinerungsprozess beeinträchtigen können. Zur Charakterisierung des Bruchverhaltens wurden Einzelkorn-Prallversuche an einer neuen Apparatur durchgeführt. Anhand der hierbei gewonnenen Parameter läßt sich die Bruchwahrscheinlichkeit der Substanzen beschreiben. Des Weiteren konnte gezeigt werden, dass ebendiese Parameter auch die Bruchfunktion, d. h. die Partikelgrößenverteilung nach der Zer-kleinerung, bestimmen, und somit eine vollständige Beschreibung des Bruch-verhaltens ermöglichen. Mittels Nanoindentation wurden mechanische Eigenschaften der Stoffe bestimmt und mit den Bruchparametern verglichen. Hierbei konnte gezeigt werden, dass ein direkter Zusammenhang zwischen der Sprödheit (d. h. dem Verhältnis der Härte zur Bruchzähigkeit) eines Stoffes und seinem Zerkleinerungsverhalten besteht.

Zur Charakterisierung der Transporteigenschaften sowie des Haft- und Anback-verhaltens wurden Scherversuche durchgeführt, und Oberflächeneigenschaften mittels Rasterkraftmikroskopie (AFM) und inverser Gaschromatographie unter-sucht. Hierbei konnten keine stoffspezifischen Eigenschaften identifiziert werden, die für eine verstärkte Partikelhaftung verantwortlich wären. Eher scheinen geometrische Faktoren wie z. B. glatte bzw. raue Oberflächen die Adhäsion zu dominieren. Generell sind die Ergebnisse aus dem AFM aber schwer auf makroskopische Prozesse wie Scher- oder Mahlversuche zu übertragen, da sich die auftretenden Beanspruchungen stark unterscheiden.

CONTENTS

INHALTSVERZEICHNIS

EINLEITUNG

Die meisten aller pharmazeutischen Produkte werden heutzutage als feste Darreichungsformen verkauft, z. B. als Tabletten, Kapseln oder Trockenpulver zur Inhalation. Wie bei vielen anderen Anwendungen auch hängen die Eigenschaften des fertigen Produkts von der Größenverteilung der Primärpartikeln ab. So erhöht sich z. B. die Bioverfügbarkeit von vielen schlechtlöslichen Wirkstoffen deutlich, wenn die Primärpartikelgröße verringert wird. In Formulierungen für die Trockenpulverinhalation muß die aktive Substanz im Größenbereich von 1-3 μm vorliegen, um bis zu den Alveolen gelangen zu können; andernfalls werden die Partikeln von den natürlichen Filtermechanismen der Lunge abgeschieden. Auch wenn eine gleichmäßige Mischung eines Wirkstoffes mit Hilfsstoffen benötigt wird (z. B. für eine Mischung zur Tablettierung), müssen die Partikeln in einem bestimmten Größenbereich vorliegen, um eine Entmischung der verschiedenen Komponenten zu vermeiden. Diese Beispiele zeigen, daß ein fundiertes Wissen über die Handhabung von Feststoffen und mechanische Verfahrenstechnik notwendig ist für die Formulierung neuer Arzneimittel. In dieser Arbeit wird der Fokus auf den Mahlprozess gelegt, der eine wichtige Methode darstellt, um Partikeln auf eine gewünschte Größe zu zerkleinern.

In der pharmazeutischen Industrie werden permanent neue Stoffe generiert. Am Anfang eines Entwicklungsprozesses ist wenig bekannt über diese neuen Substanzen. Viele chemische und physikalische Eigenschaften können in einer frühen Entwicklungsphase mit Methoden bestimmt werden, die nur eine sehr geringe Menge an Probenmaterial erfordern. Das Bulkverhaltne eines Pulvers kann jedoch erst relativ spät untersucht werden, wenn genügend Pulver zur Verfügung steht. Zum Beispiel werden die Betriebsparameter für eine Mahlung normalerweise in aufwendigen Versuchen im Labor- und im Technikums-maßstab festgelegt. Manchmal kann ein problematisches Verhalten einer Substanz erst dann entdeckt werden: manchmal kann die gewünschte Partikelgröße nicht erreicht werden aufgrund eines ungewöhnlichen

Bruchverhaltens der Substanz. Oft bleibt das Pulver auch an den Apparatewänden hängen und verstopft nach einiger Zeit die Mühle. Somit kann kein kontinuierlicher Prozess aufrechterhalten werden, und die gewünschte Partikelgröße kann ebenfalls nicht erreicht werden. Da diese Mahlversuche relativ große Mengen an Material erfordern, kann dies eine sehr kostspielige Prozedur werden, besonders im Fall von teuren pharmazeutischen Wirkstoffen.

Herangehensweise und Ziel dieser Arbeit

Die eingangs erwähnten Beobachtungen zeigen das Potential für Verbesserungen im Entwicklungsprozess: Zeit und Geld könnten gespart werden, wenn das Handlings- und Zerkleinerungsverhalten eines noch nicht getesteten Pulvers schneller und unter weniger Materialaufwand charakterisiert werden könnte. Dies wird jedoch nicht mit einem einzigen Test möglich sein: wie obige Beispiele zeigen, müssen mehrere Aspekte berücksichtigt werden: die Bruchphysik des Materials auf der einen Seite, und die Transporteigenschaften, d. h. die Partikelwechselwirkungen, auf der anderen Seite.

In dieser Arbeit werden beide Aspekte des Mahlverhaltens adressiert: die mechanischen Eigenschaften und das Bruchverhalten von pharmazeutischen Pulvern wird mittels Einzelkornzerkleinerung, Indentation und Tablettierung charakterisiert werden. Die Adhäsionseigenschaften werden untersucht werden mit Scherexperimenten und der Charakterisierung der Partikeloberflächen mittels Rasterkraftmikroskopie (AFM) und chromatographischen Methoden. Das Ziel dieser Herangehensweise ist es, diejenigen Eigenschaften zu identifizieren, die das Handling und das Zerkleinerungsverhalten eines Materials entscheidend beeinflussen, und geeignete Methoden zu finden, um diese entscheidenden Eigenschaften zu messen.

Für diese Untersuchungen wurden anfangs vier verschiedene pharmazeutische Stoffe ausgewählt: zwei aktive Wirkstoffe, von denen bereits ein problematisches Mahlverhalten wie oben beschrieben bekannt war, und zwei andere Substanzen, die als Referenz für ein unproblematisches Mahlverhalten dienen. Später wurden die vielversprechendsten Methoden auf einige weitere kristalline organische Substanzen angewendet, um eine breitere Datenbasis zu erhalten und um die Brauchbarkeit dieser Methoden zu zeigen.

1. INTRODUCTION

The majority of pharmaceutical products are nowadays sold as solid dosage forms, e. g. as tablets, capsules, or dry powder inhalation formulations, i. e. in the form of particulate material. Like in many other applications, the properties of the final product depend on the size distributions of the primary particles. For example, the bioavailability of many poorly soluble drugs is enhanced greatly if the primary particle size is decreased. For dry powder inhalation formulations, the active compound needs to be in a size range between 1-3 μm, in order to be delivered to the pulmonary alveoli; otherwise particles will be separated by the filtering mechanisms of the lung. Also, if a uniform mixing of an active compound with excipients is required (e.g. for a tabletting mixture), the particles need to be within a certain size range to avoid segregation of the different components. These examples show that a sound understanding of solids handling and mechanical process engineering is required in the formulation of a new medication. In this work, a special focus is put onto the milling process, which is an important method to reduce particulate material to a desired size.

In the pharmaceutical industry, new chemical entities are generated permanently. At the beginning of the development process, little is known about the properties of these new substances. Many chemical and physical properties can be characterised at a very early stage of development by methods that require only very small amounts of substance. The bulk properties of a powder, however, can be assessed only relatively late in the development process, when enough powder is available. For example, parameters for a milling process are usually determined in exhaustive bench and pilot scale milling trials. Some possibly problematic behaviour of a substance is only detected at this stage: sometimes, the desired particle size may not be reached because of an unusual breakage behaviour of the substance. Often, the powder sticks to the machinery walls and may block the equipment after some time; thus, a steady process cannot be maintained and the desired particle size is also not achieved. Since these milling tests require relatively large amounts of substance, this can become a very costly procedure, especially in the case of expensive pharmaceutical powders.

Approach and aim of this work

The above-mentioned observations show the potential for improvements of the development process: time and money would be saved if the handling and milling performance of a yet untested powder could be characterised more quickly and by methods that require less material. However, this will not be achieved by one single test: as the examples above show, several influences have to be considered: the breakage physics of the material on the one hand, and the transport properties, i.e. particle interactions, on the other hand.

In this work, both aspects of the milling performance are addressed: the mechanical properties and the breakage behaviour of pharmaceutical powders will be characterised by single particle comminution, indentation and tabletting experiments. The adhesive properties will be assessed by shear experiments and characterisation of the particles' surfaces by AFM and chromatographic methods. The aim of this approach is to identify those properties that are critical to a material's handling or milling behaviour and to find suitable methods to measure those critical properties.

For these studies, initially four different pharmaceutical substances were chosen: two active ingredients that are already known to exhibit a problematic milling behaviour as described above, and two other pharmaceutical substances that served as references for a non-problematic milling behaviour. Later, the most promising methods were applied to several other crystalline organic solids, in order to get a broader data basis and show the usefulness of the methods.

2. FUNDAMENTALS

2.1. Comminution

Comminution is a very old technique that has been employed since mankind began to eat cereals. Equipment for size reduction was developed already centuries and millenia ago, exploiting water, wind and steam power. Despite the long history of this technique, the scientific treatment of comminution processes is a relatively young discipline, beginning in the late 1950's (see Bernotat and Schönert, 2000).

Upon loading, particles deform elastically and plastically. A non-uniform stress field is built up, which may lead to particle breakage. Bernotat and Schönert (2000) describe many factors that influence this complex process: those involve both stressing conditions such as the energy input, kind of stress, deformation velocity or temperature, and particle properties such as size, shape, homogenity and thermo-mechanical material properties. Detailed descriptions of the correlations of above-mentioned quantities have been the subject of many studies. In the late 19th century, the earliest laws of size reduction were proposed: von Rittinger (1867) postulated that the increase in surface area is proportional to the energy consumption, which means that the energy utilisation, i. e. the ratio of increased surface area to energy consumption, is constant. Kick (1885) assumes that the strength is constant, and fracture surfaces are geometrically similar. From this it is concluded that the energy utilisation is inversely proportional to the particle size. Bond (1952) developed a relationship where the comminution work is inversely proportional to the square root of particle size, which is an approximation for the comminution in ball mills.

For a better understanding of the complex grinding process, the whole process can be divided into a machine function and a material function (Vogel and Peukert, 2003). The machine function comprises the type of mill and operational parameters, that determine the stressing conditions. The material function contains particle and material properties that are related to grinding performance, describing the response of the particles to the stressing conditions. The material function can be described by the breakage probability S and the breakage function. The breakage probability S is the statistical probability that a particle breaks at a given stress. The breakage function describes the size distribution of the comminuted material.

By dimensional reasoning, Rumpf (1973) derived the following equation to describe particle breakage in the case of particles of similar shape but different material:

$$S_v x = f\left\{\frac{E_v}{E}, \frac{E_v x}{\beta_{max}}, \frac{v_{fract}}{v_{el}}, \frac{v_d}{v_{el}}, \frac{l_i}{x}, v\right\} \quad \text{with} \quad v_{el} = \sqrt{\frac{E}{\rho}} \qquad (2.1)$$

In this equation, S_v is the newly created specific surface area, E_v the volume-specific energy supplied to the particles, E the Young modulus, β_{max} the maximum fracture energy (also called crack extension energy), v_{fract} the velocity of crack propagation, v_{el} the velocity of elastic wave propagation, v_d the velocity of deformation, l_i the length of the flaws initially present in the material, x the initial particle size, v the Poisson ratio and ρ the particle density.

In single particle experiments, well-defined stresses can be applied to particles. This method allows characterisation of the particle-specific influences on comminution, i. e. the material function. Various authors used single particle impacting for characterising different materials: Yuregir et al. (1986) and Ghadiri and Zhang (2002) used a jet impact device to study impact attrition of NaCl, KCl and MgO. Salman et al. (1995) studied particle failure of aluminium oxide under different impact angles. While those experiments were done with particles in the millimeter size range at relatively low impact speeds ($v \leq 35$ m/s), other researchers examined particles of a few 100 μm diameter at higher impact speeds: Menzel (1987) comminuted various size fractions of limestone and studied the influence of different process parameters on impact comminution. Lecoq et al. (2003) impacted different materials (glass, sand, polyamide, NaCl, Al(OH)$_3$, PMMA) to characterise the differences in breakage behaviour. Vogel and Peukert (2003) use a single particle device after Schönert and Marktscheffel (1986) to characterise the material function. Kwan et al. (2004) used single particle impact studies to characterise the milling behaviour of pharmaceutical powders.

Results from single particle experiments can be used not only for the characterisation of a material, but also as an input for the modelling of comminution processes (see e. g. Austin (1971); Müller et al. (1999);

Gommeren et al. (2000); de Vegt et al. (2005b)). However, as of now there is still no comprehensive understanding of the entire milling process, and in practice, a given milling problem is usually solved by extensive milling trials.

2.1.1. Models describing impact comminution

In the following, some approaches are explained to describe particle breakage. This gives not only an overview over the existing models but also shows the relevant quantities that are generally considered to have an influence on particle breakage.

Vogel and Peukert (2003) assumed that breakage starts from material flaws that are most stressed at the border of the contact zone during an impact. Combining Hertz analysis with Weibull statistics, they derived the following equation for the breakage probability:

$$S = 1 - exp\{-f_{Mat}kx(W_{m,kin} - W_{m,min})\} \tag{2.2}$$

In this equation, x denotes the initial particle size, k the number of impacts and $W_{m,kin}$ the specific (mass-related) kinetic impact energy of the particles. The material parameter f_{Mat} comprises the breakage-relevant particle and material properties and serves as a measure for the resistance of the material against fracture. $W_{m,min}$ can be regarded as energy threshold, it characterises the specific energy which a particle can take up without comminution. This energy threshold is size-dependent; the product of particle size and energy threshold $x \cdot W_{m,min}$ was found to be constant. For purely elastic particle stressing, f_{Mat} and $x \cdot W_{m,min}$ can be calculated using Hertz analysis. In reality, elastic-plastic behaviour occurs in most cases. Since the extent and the effect of inelastic material deformations are not known, the material parameters f_{Mat} and $x \cdot W_{m,min}$ need to be determined experimentally by single particle impact tests at different impact speeds. Having impacted a sufficient number of particles, the breakage probability can be set equal to the fraction of broken particles, which can be determined by sieving. The breakage probabilities obtained from these experiments are plotted over the impact energy, and a curve fitting using Eq. (2.2) finally yields f_{Mat} and $x \cdot W_{m,min}$. Based on Eq. (2.2), the breakage probabilities of different materials with different sizes can be described by a single mastercurve, as shown in Fig. 2.1. From their experimental data, Vogel and Peukert found that the breakage parameters seem to be inversely correlated.

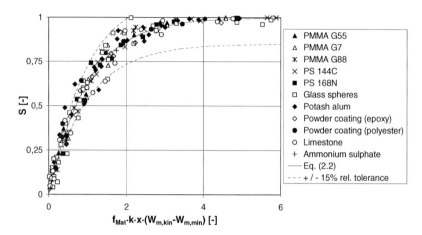

Fig. 2.1 Master curve for the breakage probability after Vogel and Peukert (2003)

De Vegt et al. (2005a) used dimensional analysis to derive the following equation describing particle breakage during milling:

$$S_i = c \frac{E_v E_{fract} \sqrt{\frac{P_y}{\rho}}}{H\sqrt{x} K_{IC}} \left(\frac{l_i}{x}\right) \tag{2.3}$$

Here, S_i denotes the breakage rate of a particle with size i per time interval; c is a constant, E_{fract} the fracture energy, P_y the yield pressure, ρ the particle density, H the hardness, x the particle size, K_{IC} the stress intensity factor and l_i the initial flaw size. While process parameters (except E_v) are comprised in the constant c, this approach also emphasises the dependency of milling performance on the material and particle properties, which de Vegt et al. estimate for different pharmaceutical substances from compression experiments and other correlations, e. g. K_{IC} is determined from the solubility parameter δ (Roberts et al., 1995).

Attrition is different from comminution in the respect that attrition happens mainly locally (e.g. at edges and corners of a particle), while in comminution, the stress and the resulting cracks penetrate the entire particle. Attrition happens already at stress levels that are much smaller than those necessary for comminution. In spite of these differences, the basis of both processes is very similar: the creation of a stress state that leads to the propagation of cracks.

Ghadiri and Zhang (2002) developed a model to describe chipping for materials having a semi-brittle failure mode. They derived the following equation:

$$\xi = \alpha \frac{\rho v^2 x H}{K_C^2} \, ,$$ (2.4)

where ξ is the fractional loss from a particle due to impact, α is a proportionality factor, ρ the density of the material, v the impact velocity, x the particle diameter, H the hardness and K_C the fracture toughness of the material. Kwan et al. (2004) employ this model also for the characterisation of particle breakage. A similar relation is derived by Gahn (1997). He quantifies the fraction of attritted volume by

$$\frac{V_a}{V} = \alpha^* \frac{\rho^{4/3} x H^{2/3}}{G \beta_{max}} v^{8/3}$$ (2.5)

where V_a/V is the fraction of attrited volume, α^* is an experimental proportionality constant, and G is the shear modulus. Assuming linear-elastic material behaviour, the product $G \cdot \beta_{max}$ is proportional to the square of the fracture toughness K_C. Both presented attrition models work well for characterising attrition of brittle and semi-brittle materials, but not for viscoelastic polymers, as was found by Frye and Peukert (2002).

The dependency of the breakage probability on the impact angle was also studied by several authors. Menzel (1987) varied the impact angle of limestone between 30° and 90° and found practically constant results for impact angles of 75° and above; at lower angles, a coarser product was obtained. Salman et al. (1995) studied oblique impacts of 5.15 mm aluminium oxide spheres. They report a decrease of the breakage probability at angles below 50°, while it varies only slightly at angles between 50° and 90°.

2.1.2. Breakage functions

Besides the breakage probability, also an expression for the breakage function is needed in order to model a milling process. The breakage function describes the size distribution of the fragments of a particle after breakage. Usually, the approaches are based on empirical equations whose parameters are correlated with operating parameters of the respective mill (e.g. Müller et al., 1999). A number of mathematical functions have proven to be useful for describing particle size distributions. Table 2-1 gives an overview over the most common ones. These are the GGS distribution (Gates-Gaudin-Schumann, also known as

7

power law distribution function), the RRSB distribution (after Rosin, Rammler, Sperling and Bennett) and the logarithmic normal distribution. The first two models are purely empirical in nature. As pointed out by Gebelein (1956), the RRSB distribution usually overestimates the number of very fine particles significantly (if the lower limit approaches 0, the number of particles becomes infinite for $n \leq 3$) and accordingly also yields unrealistic high specific surface areas. Therefore, often another parameter has to be introduced as the lower size limit of the RRSB distribution. The same problem arises with a power law distribution. For example, Vogel and Peukert (2005) use a power law distribution to describe the breakage functions of their single particle experiments. In order to reduce the high fraction of fine particles in this model, they superimpose an attenuation function that reduces the fraction of fines below a given particle size limit, which was adapted for each substance from the experimental data obtained from sieving. Using this approach in combination with the Weibull distribution for the breakage probability given above, Vogel and Peukert were able to simulate milling processes in a sieve hammer mill and an air classifier mill.

Table 2-1 Commonly used models for particle size distributions, and modifications

Name	Equation	Parameters
Power law distribution (Gates, Gaudin, Schumann)	$$Q_3 = \left(\frac{x}{x_{max}} \right)^m$$	x_{max}; m
RRSB distribution (Rosin, Rammler, Sperling, Bennet)	$$Q_3(x) = 1 - exp\left\{ -\left(\frac{x}{x'} \right)^n \right\}$$	x'; n
Logarithmic normal distribution	$$q_3(x) = \frac{1}{x\sigma\sqrt{2\pi}} exp\left\{ -\frac{1}{2}\left[\frac{ln(x/x_{50,3})}{\sigma} \right]^2 \right\}$$	σ; $x_{50,3}$
Power law combined with attenuation function	$$Q_3 = \frac{1}{2}\left[1 + tanh\left(\frac{x - x^*}{x^*} \right) \right]\left(\frac{x}{x_{max}} \right)^m$$	x^*; x_{max}; m
Truncated log-normal distribution	$$Q_3(x) = \frac{1}{\sqrt{2\pi}} \int_{-\infty}^{u} exp\left(-\frac{t^2}{2} \right) dt$$ with $u = \frac{1}{\sigma}\left(ln\frac{x}{x_{max} - x} - ln\frac{x_{50,3}}{x_{max} - x_{50,3}} \right)$	σ; $x_{50,3}$; x_{max}

Many researchers found that logarithmic normal distributions are well suited to describe particle size distributions of milled products. This type of distribution avoids the afore mentioned problems in the fine particles range. The physical meaning of log-normal distributions is also underlined by several authors. Epstein (1948) showed that the product of a milling process asymptotically reaches a log-normal distribution after a sufficient number of comminution steps. Klotz (1981) considers an infinite plate with randomly distributed microcracks under uniaxial stress and shows that the resulting fragments are predominantly convex and log-normally distributed. Gebelein (1956) shows that the logarithmic normal distribution provides a more realistic picture than the RRSB distribution in the region of very fine particles, while in the region of coarse particles, RRSB has advantages over a log-normal distribution. The latter contains no upper size limit, i. e. theoretically infinitely large particle sizes are possible (although they are very rare due to statistical reasons). Therefore, an adaption to the region of real existing sizes is recommended (see also Buss, 1976). Klotz and Schubert (1982) and Hanisch and Schubert (1984) use the superposition of several truncated logarithmic normal distributions (possessing an upper size limit) to describe the size distributions of quartzite after crushing.

However, all approaches so far remain empirical; usually, parameters are adapted to the outcome of a given milling experiment with a given material at given operating conditions. Breakage functions that possess a general applicability still have to be found.

2.1.3. Definition of breakage probability

The raw data that enter the evaluation after Vogel and Peukert are the energy input and the breakage probability. Therefore it is necessary to give an accurate definition of these quantities and consider the consequences arising from this definition.

In a strict sense, the breakage probability is - as the name says - the probability that a particle breaks when exposed to a certain stress. In order to evaluate the breakage probability according to this definition, each particle needs to be inspected individually before and after stressing. This can be done e.g. with particles in the mm-size range (Salman et al., 1995), but requires a very large effort. For finer and for irregularly shaped particles, this method becomes practically impossible to do. A more practical approach is to determine the fraction of broken particles by sieving, as was done by Vogel and Peukert (2003). All material that falls through the lower sieve mesh after stressing is regarded as broken. This implies the assumption that if a particle breaks, all fragments will be small enough to pass the lower sieve mesh. Therefore, a narrow sieve fraction has to be used as feed material. Still, if only some chipping takes place, i. e. only a few corners are abraded, the remaining core particle may stay within the non-broken fraction or may just become small enough to pass the mesh. Thus, the sieving method introduces a certain kind of differentiation between slight chipping (that will be practically ignored in the result, as only a few fragments will be in the broken fraction) and a higher extent of chipping which is treated as fracture of the particle (as all fragments will end up in the broken fraction). In this respect, the result from sieve analysis can deviate from the original definition of breakage probability; but it becomes clear that the core question actually is: when is a particle defined as broken? Can chipping be regarded the same as fracture? The sieve method answers this question as follows: a particle will be regarded as fractured if its mass loss exceeds a certain amount; in this case, even the largest fragments will have a size that can pass the lower sieve mesh.

In this work, a new method was developed for the determination of breakage probability from size distributions measured by laser diffraction. This method is validated by comparison to the results of sieve analysis, therefore, basically the same definition will apply for this method. Further implications of this new methods will be discussed when the method is explained in detail in Chapter 3.1.4.

2.1.4. Definition of impact energy

The material parameters f_{Mat} and $x \cdot W_{m,min}$, as derived by Vogel and Peukert, are valid for impact comminution on a rigid target. In practice, it has to be considered that a part of the kinetic energy is being transferred to the impact plate by elastic deformation of that plate. Therefore, it will be calculated in the following, how much of the kinetic energy of the particle is being transformed to elastic deformation energy on impact. The approach is the same as the one by Becker et al. (2001), who estimated the amount of energy that can be transferred from the grinding medium to the product in a stirred ball mill. Linear-elastic material behaviour is assumed, thus Hertz analysis can be applied.

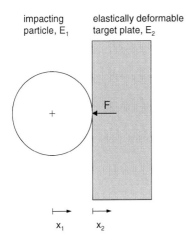

Fig. 2.2 Sphere impacting on a target plate

Consider the impact of a spherical particle on an elastically deformable plate, according to Fig. 2.2. The coordinate x_1 denotes the position of the center of the particle; chosing $x_1 = 0$ if the particle just touches the plate (no deformations), x_1 denotes the total deformation of both particle and plate. x_2 denotes the deformation of the plate, and x_1-x_2 is the deformation of the particle. The deformation is correlated to the force F between the contact partners as given by Hertz (1881):

$$x = \alpha F^{\frac{2}{3}} \tag{2.6}$$

The parameter α contains elastic properties of the material, α being reciprocal to Young's modulus E. Thus, formulating the equation of motion for the impacting particle, we obtain

$$m_{particle}\ddot{x}_1 = -\left(\frac{x_1 - x_2}{\alpha_1}\right)^{\frac{3}{2}} \tag{2.7}$$

Since the contact force F acting on the particle and the plate is the same, we can derive an expression of x_2 as a function of x_1 from

$$\left(\frac{x_1 - x_2}{\alpha_1}\right)^{\frac{3}{2}} = \left(\frac{x_2}{\alpha_2}\right)^{\frac{3}{2}} \tag{2.8}$$

Thus we can eliminate x_2 from Eq. (2.7) and obtain a solution for the maximum deformation of the particle $x_{1,max}$ by integration of this equation. The maximum of elastic energy stored in the particle can then be calculated from

$$W_{max} = \int_0^{(x_1 - x_2)_{max}} F(x_1 - x_2)\,d(x_1 - x_2)$$

(2.9)

Substituting (x_1-x_2) by $x_1 \cdot [\alpha_1/(\alpha_1 + \alpha_2)]$ (from Eq. (2.8)), this equation can be easily integrated. Inserting the solution for $x_{1,max}$ from Eq. (2.9), and recalling that $\alpha \propto \dfrac{1}{E}$ finally yields

$$W_{max} = \frac{1}{2} m_{particle} v_0^2 \left(1 + \frac{E_1}{E_2}\right)^{-1}$$

(2.10)

In terms of mass-specific energy, this can be re-written as

$$W_{m,max} = \frac{1}{2} v_0^2 \left(1 + \frac{E_1}{E_2}\right)^{-1} = W_{m,kin}\left(1 + \frac{E_1}{E_2}\right)^{-1} = W_{m,kin}^*$$

(2.11)

With this result, it becomes obvious that the maximum energy that is available for comminution depends not only on the impact velocity but also on the material properties of the impacting species and the target. Suppose, for example, an organic solid with a Young's modulus of 20 GPa (which is in the medium range of the substances studied in this work), impacting on a steel plate (E = 210 GPa): the maximum elastic energy will be only about 91% of the kinetic energy.

The higher the Young's modulus, the less energy will remain in the particle upon impact. Still, material parameters derived after Vogel and Peukert are suitable to describe comminution by impact on a steel target (or whatever material is used for the impact plate; steel was used in this work and in Vogel and Peukert's work). The approach also remains useful for a coarse ranking of different materials, especially if the Young's moduli are not known. However, if materials with very different Young's moduli are compared, the results will be skewed. Therefore, the influence of the impact plate material should be separated in order to obtain truly intrinsic material properties. This can be achieved, if we insert $W_{m,max}$ according to Eq. (2.11) instead of $W_{m,kin}$ as the kinetic energy into Eq. (2.2). The resulting equation then can be rearranged as:

$$S = 1 - exp\left\{-f_{Mat}^*\left(1+\frac{E}{E_{target}}\right)^{-1}kx\left(W_{m,kin}-\left(1+\frac{E}{E_{target}}\right)W_{m,min}^*\right)\right\}$$ (2.12)

Comparison with the "original" Eq. (2.2) shows that the corrected material parameters f_{Mat}^* and $x \cdot W_{m,min}^*$ can be easily calculated from the original parameters:

$$f_{Mat}^* = f_{Mat}\left(1+\frac{E}{E_{target}}\right)$$ (2.13)

$$xW_{m,min}^* = xW_{m,min}\left(1+\frac{E}{E_{target}}\right)^{-1}$$ (2.14)

Accordingly, $W_{m,eff}^*$ denotes the energy that is effectively available for comminution:

$$W_{m,eff}^* = W_{m,max} - \frac{xW_{m,min}^*}{x} = W_{m,kin}^* - W_{m,min}^*$$ (2.15)

2.1.5. Particle breakage and material properties

The fracturing process, as far as it is understood today, can be explained by energetic considerations. As Bernotat and Schönert (2000) explain, energy is required for the inelastic deformation in the fracture zone, for electronic processes involved in material separation and for the creation of new surfaces. Because of the high crack propagation velocity, energy can not be supplied externally, but has to be supplied from the stress field, i. e. from the energy that is elastically stored in the material.

To ensure that a crack crosses the entire particle, a minimum particle size is required. Otherwise, not enough elastic energy can be stored in the particle, and the particle will only deform plastically under stress. This is the so-called brittle-ductile transition. For a cylindrical sample, the critical size is obtained from an integral energy balance:

$$x_{crit} \geq \frac{2ER}{\sigma_B^2},$$ (2.16)

where R is termed the crack resistance, which is equivalent to β_{max}. σ_B is the breaking stress. Schönert concludes that an additional or repeated energy supply is required after the initial fracture, if finer particles are to be disintegrated.

Kendall (1978) considered the size limit for the crushing of particles and proposed a very similar equation to describe the brittle-ductile transition:

$$x_{crit} = \frac{32}{3} \frac{ER}{P_y^2}, \tag{2.17}$$

where P_y is the yield stress in a uniaxial test. A similar expression is derived by Lawn and Evans (1977) who consider the load required for the growth of pre-existing cracks in indentation experiments. They derive a critical flaw size as

$$c_{min} = 44.2 \left(\frac{K_C}{H} \right)^2 \tag{2.18}$$

where K_C is the fracture toughness, H is the hardness of the material. Note that $E \cdot R$ is proportional to the square of K_C for linear-elastic material behaviour. Note also that P_y is correlated with hardness, though not necessarily linearly (see Tabor (1986), and Eqs. (2.33) to (2.35) in Section 3.3), therefore the expressions of Kendall and Lawn and Evans are comparable to a certain extent. The results of Lawn and Evans can also be transferred to the fracture of particles: if a particle is smaller than the critical crack length obtained at critical load, it means that the particle is too small to store the energy required for fracture, and only plastic deformation will occur.

Another, also very similar expression is derived by Hagan (1979) who estimates the critical load required for crack nucleation, and the corresponding size of the cracks:

$$c_{min} = 29.5 \left(\frac{K_C}{H} \right)^2 \tag{2.19}$$

Puttick (1980) outlines a more general approach and concludes that fracture transitions depend on a material factor ER/P_y^2 (or, equivalently, $(K_C/P_y)^2$) and a test factor α which depends on the stress type and the sample geometry

$$x_{crit} = \alpha \frac{ER}{P_y^2}, \tag{2.20}$$

which is consistent with the models described before. This general correlation indicates that $(K_C/H)^2$ obviously can serve as a measure for the brittle-ductile transition. However, if a milling size limit is to be predicted for a given milling process, also the test factor α has to be known for the specific process.

Lawn and Marshall (1979) proposed the ratio of hardness and fracture toughness H/K_c as a measure for the brittleness of a material. This quantity (also called "brittleness index") contains the two competing mechanisms of plastic deformation (characterised by the hardness, i.e. "resistance to deformation") and fracture (given by the fracture toughness, "resistance to fracture"). A high brittleness index indicates a brittle substance. The brittleness seems to be related to the milling performance, as was shown by Taylor et al. (2004b), who compared brittleness indices of five pharmaceutical compounds -measured by nanoindentation- to the size reduction observed in milling experiments.

2.2. Nanoindentation

Hardness is generally defined as the resistance of a material against plastic deformation when penetrated by another body. The hardness of a material can be characterised by pressing a rigid body of defined geometry into the surface of the tested material. The hardness can be calculated as

$$H = \frac{P}{A} \tag{2.21}$$

where P is the applied force and A ist the area of the permanent impression after the test, which can be determined from the dimensions of the impression and the geometry of the indentation tip. Many hardness testing methods have been developed that work according to this principle, which vary in the geometry and the material of the employed tip. Common testing methods are Brinell (using a spherical indenter), Vickers, Knoop (pyramidal indenters) and Rockwell (conical indenter). Because the impression area is only determined after the applied force has been removed, elastic components of the deformation can not be measured with these methods.

Other techniques have been developed that analyse the load-displacement curves obtained during the experiment. The most commonly used method nowadays is the one by Oliver and Pharr (1992), which is based on an approach by Doerner and Nix (1986) and will be explained in the following. One advantage of this method is that also elastic properties can be measured. No optical determination of the impression size is necessary. This enables hardness testing also in cases when only very small indents can be made (e.g. on thin films or small particles), where the accuracy of optical methods would be limited.

The reduced modulus E_r is given by

$$\frac{1}{E_r} = \frac{\left(1 - v^2\right)}{E} + \frac{\left(1 - v_i^2\right)}{E_i} \tag{2.22}$$

where E and v are the Young's modulus and the Poisson ratio of the substrate. The quantities with the subscript "i" refer to the properties of the indenter tip. The reduced modulus can be calculated from the initial slope of the unloading curve (see Fig. 2.3) by

$$E_r = \frac{1}{2}\sqrt{\frac{\pi}{A}} \frac{1}{\beta} \frac{dP}{dh} \tag{2.23}$$

P is the applied load, dP/dh is the slope at the beginning of the unloading curve as depicted in Fig. 2.3, A is the contact area, β is a geometric correction factor (β = 1.034 for a Berkovich indenter, which is used in the experiments presented in this work). The hardness is obtained from Eq. (2.21). It should be noted that in nanoindentation testing, the projected area of impression is used for the calculation. The thus determined mean contact pressure is also called the Meyer hardness, in contrast to the Vickers hardness that is determined using the actual

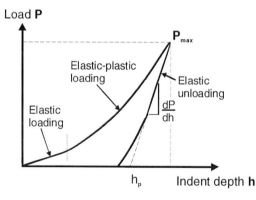

Fig. 2.3 Load over indent depth during an indentation experiment

surface area of the impression. For an ideal Berkovich indenter, $A = 24.5 h_p^2$, h_p being the so-called "plastic depth", which can be obtained from the load-displacement curve according to Fig. 2.3.

Palmqvist (1957, 1962) recognised that the work necessary to form cracks during indentation and the length of these cracks can be used as a measure for the fracture toughness of the material. Since then, many models for the evaluation of fracture toughness from indentation data have been developed and refined for ceramic materials. Ponton and Rawlings (1989a, 1989b) give an overview over 19 different model equations. Most of these models are based on the assumption of a half-penny crack geometry, some models assume a Palmqvist crack geometry. The difference of these crack systems is depicted in Fig. 2.4. Palmqvist cracks are sole radial cracks that propagate from the indent corners, while half-penny cracks are formed underneath and around the indent. It is believed that the formation of one of the crack systems is not only dependent on the material, but also depends on the load level: while at low loads, mainly Palmqvist cracks are generated, the crack system changes to half-penny type at higher loads. The fracture mechanical modelling of half-penny cracks leads to the following relationship that is required to be valid in many of the fracture toughness models:

$$\frac{P}{c^{3/2}} = \text{const.} \tag{2.24}$$

where c is the so-called crack radius which is the sum of crack length l and indent diagonal a. In contrast, a Palmqvist crack is usually described by the condition

$$\left(\frac{P}{l}\right)^{1/2} = \text{const.} \tag{2.25}$$

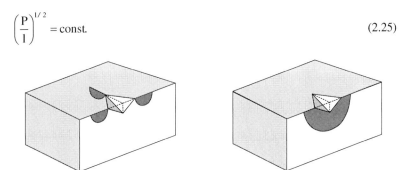

Fig. 2.4 Crack systems for Vickers indentation: left: Palmqvist cracks; right: half-penny cracks (after Fischer-Cripps, 2004)

A closer look shall be taken at the works of Evans and Charles (1976). For their experimental data, which include several ceramics covering a wide range of hardness and fracture toughness, they find a dependency in the form of

$$\frac{K_c}{H \cdot a^{1/2}} = g_1\left(\frac{c}{a}\right) \cdot g_2\left(\frac{E}{H}\right) \tag{2.26}$$

where g_1 and g_2 are independent dimensionless functions. A suitable simple form for g_2 was found to be

$$g_2\left(\frac{E}{H}\right) = \left(\frac{E}{H}\right)^{2/5} \tag{2.27}$$

For large c/a ratios, Evans and Charles derived an exponent for c/a of -3/2. However, deviations from that correlation can be seen for low c/a values in a data plot of $\log\{K_c/(H \cdot a^{1/2}) \cdot (H/E)^{2/5}\}$ versus $\log(c/a)$. Therefore a boundary condition is given for the Evans-Charles model as c/a > 2.5. Similar boundary conditions are obtained for most of the models collected by Ponton and Rawlings.

Evans (1979) applied a polynomial fit after Davis to the Evans and Charles data and thus obtained a very good correlation for the entire data range, which covered c/a values ranging from ≈ 1.5 to 6. The resulting equation for the fracture toughness can be written as

$$K_c = 0.6305 \cdot E^{0.4} P^{0.6} 10^F a^{-0.7} \tag{2.28}$$

with $\quad F = -1.59 - 0.34 \cdot B - 2.02 \cdot B^2 + 11.23 \cdot B^3 - 24.97 \cdot B^4 + 16.32 \cdot B^5$

and $\quad B = log\left(\frac{c}{a}\right)$

Note that in this final form, H was replaced by the correlation $H = 0.4636 \cdot P/a^2$, P being the applied load (this correlation is valid for the geometry of Vickers indentation). It should also be noted that the lower c/a values in the Evans and Charles data correspond to those materials with lower hardness and brittleness. Therefore, we decided to use this Evans-Davis equation (Eq. (2.28)) for our evaluation, because it can provide a good quantitative correlation also for softer materials and lower c/a values.

Various authors (Duncan-Hewitt and Weatherly, 1989; Prasad et al., 2001; Taylor et al., 2004a) used the evaluation of cracks generated during indentation for the characterisation of pharmaceutical substances. In these works, it is generally found that the condition $P/c^{3/2}$ = const. is valid also for these materials, which suggests that the models by Ponton and Rawlings can also be applied to pharmaceutical substances. However, different authors used different models, and it is not clear up to now, how well data can be compared when derived from different models. In this work, the Evans-Davis equation will be used for the evaluation of fracture toughness. A detailed comparison of different models will be given in the results section of this work.

As already mentioned earlier, the "brittleness index" is defined as the ratio of hardness to fracture toughness:

$$BI = \frac{H}{K_c} \tag{2.29}$$

This index contains the two competing mechanisms of plastic deformation (hardness) and fracture (fracture toughness). A high brittleness index indicates a brittle substance, and seems to be related to the milling performance, as was shown by Taylor et al. (2004b).

2.3. Tabletting

Powder compaction is another process that is dependent on flow behaviour and fracture behaviour of the particles involved. Therefore, also tabletting experiments were conducted in order to check whether the properties of the substances measured by nanoindentation are reflected in their compression profiles. This would give the possibility to gain an insight into mechanical properties of materials by a relatively simple method.

The most commonly used method in the pharmaceutical industry for describing a compaction process is the analysis after Heckel (1961a): He considers the reduction of pore volume in a powder bed during compaction as analogous to a first-order chemical reaction:

$$\frac{dD}{dP} \propto (1-D) \tag{2.30}$$

where D is the relative density, i.e. $D = 1-\varepsilon$, ε being the porosity, P being the applied pressure. Integration of this equation finally yields

$$ln\left(\frac{1}{1-D}\right) = KP + ln\left(\frac{1}{1-D_0}\right) \qquad (2.31)$$

or

$$-ln(\varepsilon) = KP - ln(\varepsilon_0) \qquad (2.31b)$$

where K is a constant. Plotting experimentally obtained compression data, usually a curved plot is obtained as shown in Fig. 2.5. This plot can be divided into three zones: the linear region of the plot (Zone 2) represents the region of plastic deformation, where Eq. (2.31) is applicable. It is generally agreed that the curvature in Zone 1 (at low pressure) is caused by particle rearrangement processes and fragmentation in the die. The reasons for the curvature in Zone 3 are not fully understood yet. Work hardening and polymorphic transitions at high pressures have been suggested (Chan and Doelker, 1985; Pedersen and Kristensen, 1994). Also, elastic deformation and resulting small errors in determining the punch displacements are discussed (Gabaude et al., 1999; Sun and Grant, 2001): in this region of high compression, a small deviation of ε (or D) has a strong influence of $ln(\varepsilon)$ and thus on the Heckel plot. Sun and Grant (2001) therefore recommend to be cautious when interpreting data with porosities below $\varepsilon = 0.005$.

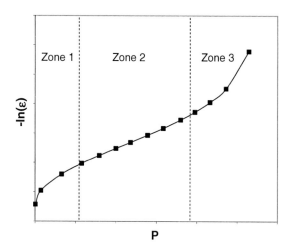

Fig. 2.5 Scheme of a typical Heckel plot

Heckel (1961b) found that the reciprocal of the slope K of the linear portion of the Heckel plot is equal to three times the yield strength σ_0 of the material. The yield strength equals one third of the mean yield pressure P_y (Hencky, 1923). Therefore,

$$\frac{1}{K} = P_y = 3\,\sigma_0. \tag{2.32}$$

The yield pressure P_y is the stress at which plastic flow starts under uniaxial tension or compression (Tabor, 1986). The indentation hardness is related to this pressure, depending on the ratio E/P_y. Tabor (1986) gives an overview as follows:

For $E/P_y < 10$: (e. g. polymers) $H \approx 1.1 P_y$ $\tag{2.33}$

For $E/P_y > 100$ (e.g. most metals): $H \approx 3 P_y$ $\tag{2.34}$

In between: $$\frac{H}{P_y} = \frac{2}{3}\left(1 + ln\frac{E \cdot cot\,\theta}{3\,P_y}\right), \tag{2.35}$$

where θ is the semi-apical angle, i.e. the included half-angle of the indenter ($cot\,\theta = 0.4$ for a Vickers pyramid). From these relation to plastic material properties, the slope K can be used as a measure for the ability of a powder to deform plastically.

There are two methods to collect compression data: in-die (also called "at-pressure") and out-of-die (also called "zero-pressure"). In "in-die" measurements, data are collected during compression, i.e. the height of the powder is measured while the compression force is still present. In "out-of-die" experiments, the tablet height is measured after the compression, i.e. when the mechanical stress - and with it the elastic, reversible deformation - is removed from the tablet. Therefore, each measured point on an "out-of-die" Heckel curve represents one tablet pressed, while "in-die", a curve can be obtained from one single compression. It should be expected that reliable values for the yield strength are obtained from an "out-of-die" Heckel analysis, since elastic deformations are excluded (Sun and Grant, 2001). Still, the "in-die" method is widely in use, probably because it allows a very fast and easy collection of data.

Paronen and Juslin (1983) and Paronen (1986) compared data from "in-die" and "out-of-die" measurements to describe the elastic properties of a powder. They introduce an elasticity parameter K_{et}, which is the reciprocal of the difference of the slopes of the Heckel plots obtained "in-die" and "out-of-die":

$$K_{et} = \frac{1}{K_{in-die} - K_{out-of-die}} \qquad (2.36)$$

It is observed in many studies, that the results of compression experiments not only depend on the applied method ("in-die" or "out-of-die"), but also on the experimental parameters, such as the die geometry and the punch speed, or the primary particle size (e.g. Sonnergaard, 1999). Therefore, care has to be taken when comparing data from literature.

In addition to the Heckel analysis, also a work of compression can be calculated by integrating the force-distance curve obtained during compression:

$$W_c = -\int_{h_0}^{h} P dh \qquad (2.37)$$

where h_0 represents the initial height of the powder bed, h the height of the bed during compression.

2.4. Particle interactions

In particle processing, interactions of particles with other particles are of great importance. Examples for such process steps are agglomeration, sintering, filtration, mixing, conveying, classifing or dispersing. Also a process step such as milling can be affected by particle interactions, namely by their influence on the transport properties: as explained before, it is often observed that a material scales at the equipment walls and thus interrupts conveying, and a steady process cannot be kept running. Therefore, knowledge and control of particle interactions are of great interest in mechanical process engineering.

Different experimental approaches are taken for the characterisation of particle interactions: adhesion of single particles can be measured with the ultracentrifugal technique (Krupp, 1967), or directly with micro-cantilever systems, as will be explained in the following subsection. While these methods study particle interactions at single particle level on a microscopic scale, the macroscopic behaviour of a powder can be characterised by bulk methods, using various types of shear testing devices, such as the Jenike shear tester (Jenike, 1964) or the ring shear tester (Schulze, 1994). Shear testing will be explained in more detail later in this chapter.

2.4.1. Single particle adhesion

Fig. 2.6 shows the main influence factors on particle adhesion (see also Götzinger, 2005). Dispersive forces, also known as van-der-Waals forces, are always present. Hamaker (1937) deduced an expression for the van-der-Waals force between spherical particles. With decreasing particle size, van-der-Waals forces become increasingly important. Adhesion is also influenced by geometric factors that affect the contact area: the surface roughness usually decreases the adhesion force (Löffler and Raasch, 1992), because the volume in the contact zone is reduced and the contact area is decreased, while deformations in the contact zone lead to an increase of adhesion. Also, polar interactions may occur due to molecular groups on the surfaces. Adsorbed surface layers can lead to capillary adhesion forces.

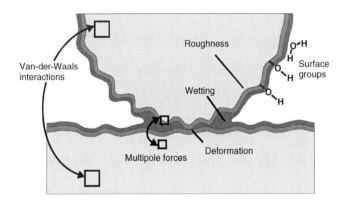

Fig. 2.6 Main influence factors on particle adhesion (from Götzinger, 2005)

2.4.2. Adhesion models

Hamaker (1937) derived an expression for the van-der-Waals adhesion force between two spherical particles:

$$F_{adh} = \frac{A}{6z_0^2}\left(\frac{R_1 R_2}{R_1 + R_2}\right) \tag{2.38}$$

where z_0 is the minimum contact distance which is given as $z_0 = 1.65 \cdot 10^{-10}$m by Israelachvili (1992). A is the Hamaker constant that contains the material-

23

specific properties. The geometry is considered by the radii R_1 and R_2 of the particles and by the minimum contact distance. This model does not take into account the deformation of the contacting particles. Other models were developed that consider deformations, such as the JKR model (after **Johnson**, **Kendall** and **Roberts**, 1971) and the DMT model (after **Derjaguin**, **Muller** and **Toporov**, 1975).

Johnson et al. (1971) assume an equilibrium between mechanical energy, elastically stored energy and the surface energy. Applying the Hertz equations, they derive an adhesion force for a spherical particle to a plane:

$$F_{adh} = 3\pi\gamma R \tag{2.39}$$

where γ is the surface energy, R is the radius of the particle. Derjaguin et al. (1975) developed another model, taking into account interaction forces outside the contact radius. They obtain

$$F_{adh} = 4\pi\gamma R \tag{2.40}$$

Maugis (1992) suggests that the JKR and the DMT model are limiting cases of a more general model. He introduces a parameter λ that takes into account the elastic and plastic material properties:

$$\lambda = \frac{2\sigma_0}{\sqrt[3]{\frac{32\pi\gamma E^2}{9R(1-\nu^2)^2}}}, \tag{2.41}$$

σ_0 being the yield strength, E being the Young modulus, ν being the Poisson ratio. For $\lambda \to \infty$, the JKR model is applicable, i.e. for soft materials with large radius and small surface energies, while for $\lambda \to 0$, the DMT model applies.

2.4.3. Influence of surface roughness

Experiments show that measured adhesion forces are often below the values predicted by the models described above. Also, large variations in adhesion forces are observed. This is attributed to the roughness of the surfaces. Rumpf (1975) proposes to model surface roughnesses as asperities with a half-spherical shape with radius r. For the contact problem of a smooth sphere in contact with a rough plane (see Fig. 2.7), the following equation is obtained (Rabinovich et al., 2000):

$$F_{adh} = \frac{A}{6z_0^2} \left[\frac{rR}{r+R} + \frac{R}{\left(1+\dfrac{r}{z_0}\right)^2} \right] \qquad (2.42)$$

This model was modified by Rabinovich et al. (2000). They replace the radius r of this asperity by an expression that contains the *rms* roughness of the surface, a quantity that it easy to measure. They obtain:

$$F_{adh} = \frac{AR}{6z_0^2} \left[\frac{1}{1+\dfrac{R}{1{,}48\,rms}} + \frac{1}{\left(1+\dfrac{1.48\,rms}{z_0}\right)^2} \right] \qquad (2.43)$$

Götzinger and Peukert (2004) showed that the measured adhesion force depends on the local roughness at the points of contact, and therefore, force distributions are obtained if adhesion forces are measured at different points of a surface. The shape of the distribution curves depends on the roughness profiles of both adhesion partners.

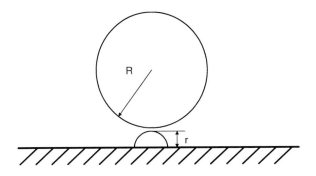

Fig. 2.7 Roughness model after Rumpf (Rabinovich et al., 2000)

2.4.4. Measurement techniques

Since more than 40 years, centrifugal techniques such as described by Krupp (1967) are used for the characterisation of particle adhesion. This method bears the advantage that many particles can be studied with one experiment; this allows reasonable statistical evaluations of the results, which is especially useful for irregularly shaped particles. However, due to limitations in the experimental setup (namely the rotational speed), this technique is confined to particles of only a few micrometers in size. Centrifugation has also been applied to pharmaceutical powders by Podzeck et al. (1994, 1995 and 1996) who studied the adhesion of pharmaceutical powders on tablets of the same substance (so-called "autoadhesion") and on wall substrates.

Another method to study single particle adhesion is by the use of micro-cantilever systems. With the development of the atomic force microscope (AFM), this method experienced a considerable boom. Decker et al. (1991) were among the first ones to glue a colloidal particle to the tip of an AFM cantilever to study the force between this particle and a planar surface in aqueous environment. This so called "colloidal probe technique" has meanwhile become a well-established tool for the study of adhesion forces of single particles. The meaning of this technique and its capabilities is underlined in an exhaustive review by Kappl and Butt (2002). They address basic questions such as the influence of applied load and loading time on adhesion, and the influence of humidity, surface roughness or surface coverings with polymer (the latter topic being of interest mainly in aqueous systems). They also give an overview over the fields of application of the technique, such as flotation (where the adhesion of the target substance to air bubbles is of interest), surface coatings, dry powder inhalation systems (here, the main interest lies in the adhesion of active drug substance to the carrier particles), paper making, printing and surface polishing.

2.4.5. Applications in the pharmaceutical area

As already mentioned, the field of dry powder inhalation (DPI) is an area where particle adhesion plays a decisive role. The active pharmaceutical ingredient (API) in the size range of a few micrometers is bound to a larger carrier particle (e.g. lactose). Upon inhalation, the carrier particles are easily dispersed in the air stream; ideally, the API particles are now also released from the carrier particles and may be delivered to the alveoli; for an optimum formulation, the adhesion of the API should be controlled in that way that the particles rather stick to the carrier particles than agglomerate with themselves, but are fixed to the carrier loosely enough to be easily redispersed.

Many studies were performed in this field: Willing et al. (2000, 2001) studied the adhesion of lactose carrier particles on commercial gelatin capsules. They found not only differences in adhesion on different kinds of capsules, but also spatial variations of adhesion forces that were attributed to surface inhomogeneities of the capsules, which were the consequence of different processing conditions for the capsules. Louey et al. (2001) studied the adhesion of a 10 µm silica sphere on lactose carrier particles obtained from different manufacturers. They found a logarithmic normal distribution of the adhesion forces with standard deviations between 1.5 and 2.3, and were able to detect differences between different lactose specifications. Sindel and Zimmermann (2001) assessed the adhesion of lactose particles on lactose tablets. Tsukada et al. (2004) compared the adhesion of DPI-sized drug particles on lactose tablets to the adhesion on a steel plate which served as a reference for a wall material of processing equipment. Young et al. (2002) made similar studies characterising the adhesion behaviour of salbutamol sulfate particles on salbutamol sulfate tablets at different relative humidities. In the same research group, Price et al. (2002) began to use purpose-grown crystals with atomically flat surfaces as substrates, instead of powder compacts. Thus the variations in contact area between colloidal probe and substrate could be reduced. In this work, the adhesion of salbutamol sulfate and budesonide on α-lactose monohydrate was tested. Begat et al. (2004) developed a method to characterise whether a substance adheres preferably to a given carrier substance or rather adheres to particles of its own kind (i.e. it agglomerates), the so-called "**c**ohesive-**a**dhesive **b**alance" (CAB). This CAB is basically the ratio of the adhesion force on a foreign material to the adhesion force on the same material and can be used to judge whether a system is likely to form a homogenous blend or rather tends to segregate upon handling and processing.

2.4.6. Bulk powder behaviour and flowability

Various types of shear testing devices are available for the measurement of flow properties of bulk solids. An overview is given by Schwedes and Schulze (1990). The most well-known device nowadays is probably still the Jenike shear tester (Jenike, 1964). In this device, the relative motion of two parts of a solid material is realised. A vertical normal force N and a horizontal shear force S are applied. Dividing the forces by the cross-sectional area gives the normal stress σ and the shear stress τ. Shear experiments at different normal loads, on a series of identically preconsolidated samples, gives a maximum shear stress for every normal stress. These data can be plotted in a σ-τ-diagram, the resulting curve is

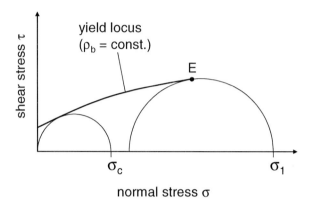

Fig. 2.8 Yield locus

called the "yield locus" (see Fig. 2.8) which refers to one value of the bulk density ρ_b. Each yield locus terminates at a point E which characterises the steady state flow of the solid.

Two Mohr cycles are shown in Fig. 2.8. The major principal stresses of these cycles are characteristic for a particular yield locus: σ_1 is the major principal stress for steady flow, called the major consolidation stress, σ_c is the unconfined yield strength. The relationship between σ_1 and σ_c can be illustrated as follows: a sample is consolidated under a normal stress σ_1 (see Fig. 2.9) After the consolidation, the sample is loaded with an increasing stress until failure occurs, i.e. the consolidated bed starts to flow. The stress at failure is the unconfined yield strength σ_c.

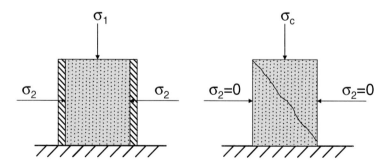

Fig. 2.9 Relation between consolidation stress and unconfined yield strength

The flowability ff_c of a powder is defined as the ratio of consolidation stress and unconfined yield strength:

$$ff_c = \frac{\sigma_1}{\sigma_c} \qquad (2.44)$$

The flowability of a powder is not a constant, but depends on the bulk density, i. e. on the preconsolidation stress. This has to be kept in mind when different powders are compared. The higher the ff_c, the better is the flow behaviour of the powder. Generally, the flowability can be divided into the following classes:

$ff_c > 10$:   free flowing

$4 < ff_c < 10$:   easy flowing

$2 < ff_c < 4$:   cohesive

$1 < ff_c < 2$:   very cohesive

$ff_c < 1$:   non-flowing

In wall friction measurements, the base of the shear cell is replaced by a sample of wall material. A wall yield locus in a diagram of σ_w vs. τ_w can be obtained in the same way as described above. Often, the wall yield locus is a straight line through the origin. Then, the wall friction angle φ is given by

$$\varphi = arctan\left(\frac{\tau_w}{\sigma_w}\right) \qquad (2.45)$$

The wall friction angle φ is a measure for wall friction. The larger φ, the higher is the friction on the wall.

2.5. Inverse Gas Chromatography (iGC)

Inverse Gas Chromatography (iGC) has become a widely used technique for the characterisation of surface properties of pharmaceutical powders. Chromatography in general works according to the following principle: An inert carrier gas (mobile phase) is flowing through a packed powder bed (stationary phase). A sample in gaseous state is added to the mobile phase. The gas molecules interact with the solid surface; depending on the strength of these interactions, different gas molecules require different times to pass the entire powder bed. Hence, gas mixtures can be separated. In analytical gas chromatography, the retention times on a known column are used to separate and identify the gaseous substances. Inverse gas chromatography uses the retention times of different gases with known properties (so-called "molecular probes") to characterise the surface of the solid material under study, which is packed into the column.

From the retention times of a series of unbranched alkanes, the dispersive component of the surface energy, γ_s^d, can be determined. The net retention volume V_N is calculated according to Schultz et al. (1987):

$$V_N = j \dot{V}(t_R - t_0) \tag{2.46}$$

t_R is the retention time of the probe, t_0 is the mobile phase hold up, or "dead time", measured with a non-adsorbing probe (methane). $\dot{V}$ is the flow rate, j the James-Martin correction factor which is necessary to account for the pressure drop over the column due to the compressibility of the gas. It is given by:

$$j = \frac{3}{2} \frac{(p_{in}/p_{out})^2 - 1}{(p_{in}/p_{out})^3 - 1} \tag{2.47}$$

where p_{in} and p_{out} are the the pressures at the inlet and the outlet of the column. The free enthalpy of adsorption ΔG^0 of n-alkanes is given by

$$\Delta G^0 = RT \ln V_N + C \tag{2.48}$$

where R is the universal gas constant, T is the temperature, C is a constant that depends on the chosen reference state. The free enthalpy of adsorption is related to the energy of adhesion W_A between probe and solid by

$$\Delta G^0 = N a W_A \tag{2.49}$$

where N is Avogadro's number and a the surface area of the probe molucule. According to Fowkes (1964), in the case of dispersive interactions as e.g. for n-alkanes, the energy of adhesion is given by

$$W_A = 2\sqrt{\gamma_S^d \gamma_L^d} \tag{2.50}$$

where γ_S^d and γ_L^d are the dispersive components of surface energy of the substrate and the adsorbate. Combining Eqs. (2.48) - (2.50) leads to the relationship

$$RT \ln V_N + C = 2 N_A a \sqrt{\gamma_S^d \gamma_L^d} \tag{2.51}$$

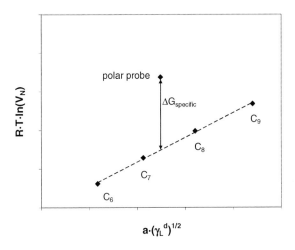

Fig. 2.10 Schematic representation of the alkane line and specific interactions of polar probe molecules

By plotting $RT \cdot ln(V_N)$ vs. $a\sqrt{\gamma_L^d}$ for a homologous series of alkanes, a straight line is obtained, as indicated in Fig. 2.10. From the slope of this line, the dispersive component of surface energy can be calculated. Polar probes lie above this alkane line, and a specific enthalpy of adsorption can be calculated from the distance between the enthalpy of the polar probe and the alkane line.

Adsorption equilibria at low pressures can be described by Henry's law:

$$n_{ads} = Hp \tag{2.52}$$

where n_{ads} is the loading of the adsorbent with adsorbate, p the pressure and H the Henry coefficient. The higher the Henry coefficient, the more adsorbate is adsorbed. Hence, the Henry coefficient can be used to characterise the adsorptive properties of a surface. The Henry coefficient can be calculated from the net retention volume obtained from iGC:

$$H_{S,BET} = \frac{H}{S_{BET}} = \frac{V_N}{S_{BET} m_S RT} \tag{2.53}$$

$H_{S,BET}$ is the surface specific Henry coefficient, H the mass specific Henry coefficient, S_{BET} the BET surface of the solid and m_S the mass of the solid. For a more detailed discussion about Henry coefficients and their determination see Maurer (2000).

3. EXPERIMENTAL SETUP

3.1. Impact comminution

3.1.1. Single particle impact device after Schönert

Fig. 3.1 shows the single particle impact device after Schönert and Marktscheffel (1986). With a vibrational feeder, particles are fed individually into the center of the rotor. Particle-particle interactions in the rotor are excluded due to the slow manner of feeding. The particles are accelerated towards the outer diameter of the rotor along one of four radial channels by centrifugal forces. The radial component of the partical velocity is equal to the tangential component. Thus, particles leave the rotor at an angle of 45°. In the milling chamber, particles hit the impact ring which is shaped in a way that always an impact angle of 90° is achieved. The entire device is evacuated, therefore no disturbing air flows can occur that might slow the particles down or deflect them from their path. Therefore, the impact velocity is equal to the velocity at which the particles leave the rotor and can be calculated from the revolution number n of the rotor:

$$v_{impact} = \sqrt{2}\pi d_R n \tag{3.1}$$

where d_R is the diameter of the rotor. The specific impact energy is determined from the impact velocity:

$$W_{m,kin} = \frac{1}{2} v_{impact}^2 \tag{3.2}$$

After impact, particles and fragments fall onto the bottom of the milling chamber and can be collected after the experiment for analysis. The setup of this apparatus enables well-defined stressing conditions: one single impact at 90° with a known impact velocity.

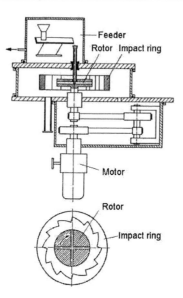

Fig. 3.1 Single particle impact device by Schönert and Marktscheffel (1986)

3.1.2. Single particle air jet device

One drawback of the Schönert device is the experimental time that increases with decreasing particle size: Since no air flow is present in this device, the dispersion of the particles can only be achieved during feeding. This means, the smaller the particles, the slower the feed rate needs to be. Additionally, particles increasingly tend to stick together with smaller particle sizes. This can, in the worst case, make a proper characterisation impossible, because if too many particles are fed into the rotor at the same time, particles will interact with each other, causing undefined stress conditions. Thus, the lower limit for this device in terms of particle size is about 100 µm for particles which are not charged electrostatically, and even higher for organic particles which are prone to triboelectric charging.

Because of this drawback, another impact device was constructed that uses an air flow to disperse particles and accelerate them towards an impact plate. This device has been constructed especially to study impact comminution of finer powders. A description of this device has been published by Meier et al. (2008), and will be given in more detail in this section. Similar devices that work after the same principle have been described in the literature (Yuregir et al., 1986; Menzel, 1987; Salman et al., 1995; Lecoq et al., 2003).

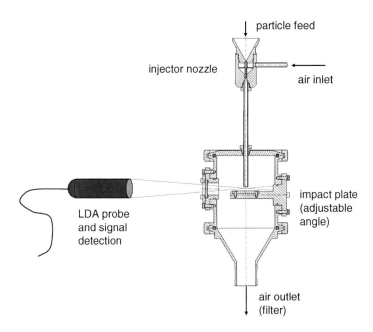

Fig. 3.2 Setup of the new impact jet device

The scheme of the air jet device is shown in Fig. 3.2. With a vibrational feeder set to a small feeding rate, particles are fed into a Venturi tube where they are dispersed in a stream of compressed air. It is not necessary to feed the material as single particles, since also loose agglomerates can be dispersed over the tube diameter. If the feed rate is kept small, the particle concentration can be regarded as sufficiently low to neglect particle-particle interactions during acceleration and impact. Typically, the feed rate in the experiments was about 20 g/h, resulting in solid to air mass flow ratios lower than 0.01. Lecoq et al. (2003) pointed out that at solid to air flow rate ratios of less than 0.1, experiments are comparable with single particle impacts. The particles are accelerated along a tube (400 mm long, 6 mm inner diameter) and hit an impact plate made of hardened steel. The impact angle in the device can be varied, the impact velocity is controlled by the air flow. The impact speed of the particles is determined from Laser-Doppler anemometry. In order to avoid further comminution when particles hit the walls of the milling chamber, the walls are covered with 0.5 mm thick PVC foil. After the impact, the material is collected in a filter sack at the bottom of the impact chamber.

Initially, experiments were conducted using a standard nozzle after ISO 5011 for dispersion of the particles. With this design, particles were fed into a horizontal air stream, and a piece of tubing was necessary to lead this stream into the vertical acceleration tube. Thus, particles are forced to change their direction of flight twice, and there is an increased risk that attrition or comminution takes place already during dispersion of the particles. Therefore, a new nozzle was constructed later according to the design that can be seen in Fig. 3.2. Now, particles can fall in a straight line down into the acceleration tube towards the impact plate. Not only the breakage in the nozzle could be reduced noticeably by this construction, also the pressure drop over the nozzle was significantly smaller. This allowed the use of a longer acceleration tube in order to gain a higher particle velocity. For experiments with the new nozzle design, a tube length of 600 mm was used.

3.1.3. Measuring the impact velocity by LDA

Laser-Doppler Anemometry (LDA) is a well-established optical method for the determination of particle velocities in a flow. It is based on a frequency shift of monochromatic light due to the motion of a particle (similar to the acoustic Doppler effect). Because LDA is a non-invasive method, the flow field is not influenced by the measurement. Only an optical access to the system has to be installed.

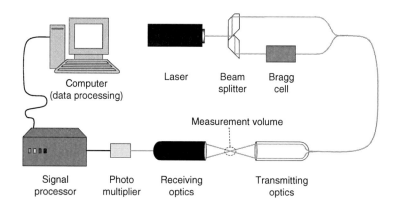

Fig. 3.3 Setup of a dual beam laser-Doppler anemometer system

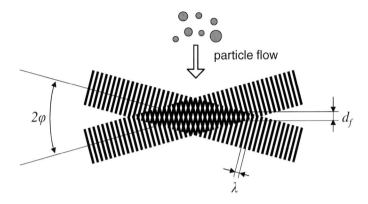

Fig. 3.4 Derivation of the fringe model by superposition of two laser beams

The scheme of an LDA system is depicted in Fig. 3.3. First, a laser beam is split into two separate beams of the same frequency. These coherent beams are then intersected at a defined angle. The interference of these beams yields a pattern of regularly varying light intensities, the so-called interference fringes (see Fig. 3.4). The fringe distance d_f is determined from the angle of intersection, 2φ, and the wavelength λ of the light:

$$d_f = \frac{\lambda}{2 \sin \varphi} \tag{3.3}$$

If a particle is moving through the measurement volume, it will scatter light from the two beams. Depending on its exact position, the intensity of the scattered light varies. From the measured frequency f of these variations, the velocity v_p of the particles can be calculated by:

$$v_p = d_f f \tag{3.4}$$

The best signal quality is generally obtained with particles in the size range around the fringe size and smaller. However, good signal qualities are also obtained for larger particles. This can be explained by a more detailed approach based on Mie's theory. Further information on this can be found in Eliasson and Dändliker (1974), Durst et al. (1981) and Ruck (1987).

The system as described above would be insensitive to the flight direction of the particles. Particles coming from opposite directions with the same velocity will cause the same frequency of scattered light. To resolve this, the frequency of one of the beams is shifted in a Bragg cell (usually by 40 kHz). As a result of this frequency shift, the fringe pattern will be moving, and the frequency of the scattered light will be different for particles coming from different directions because the relative velocity between particle and fringe pattern has changed.

The LDA system used here consists of an Innova 70 argon ion laser (Coherent Inc., Santa Clara, California, USA), and a Dantec 60X FiberFlow optical system (Dantec Dynamics A/S, Skovlunde, Denmark). Data were evaluated using the software flowPOINT by ILA GmbH, Jülich, Germany. Measurements were done using green light with a wave length of 514.5 nm, beams were intersected at $2\varphi = 5.5°$, resulting in a fringe distance of 5.4219 µm. Signals were collected in backscattering direction, i.e. the receiving optics are integrated in the transmitting probe.

To measure the particle velocity properly, the particles have to cross the measurement volume perpendicular to the axis of the collecting lens and in the plane of the two laser beams. Due to the geometric setup of the impact device, it is not possible to locate the measurement volume directly onto the surface of the impact plate in the correct orientation. Therefore, the real impact velocity is not measured, but rather the particle velocity about 2 mm above the plate. It will be shown in Chapter 4.2.2 that the error arising from this is neglegible for particles larger than several 10 µm.

3.1.4. Determination of breakage probability

The breakage probability is determined in this work from the particle size distributions before and after the experiments. Two different methods are employed to measure the size distributions: sieving and laser light diffraction.

From sieving, the breakage probability is readily determined: the material before milling consists of a single sieve fraction. After comminution, the sample is sieved again, and all fragments that are now passing through the lower mesh of the initial fraction represent the broken fraction. For a large number of impacted particles, this broken fraction approaches the statistical breakage probability. Assuming spherical particles of 500 µm diameter with a density of 2000 kg/m³, a sample of 1 g would contain 7639 particles. The experiments conducted in this work were done with particles smaller than this size and with a lower density; sample sizes around 10-15 g were used. Therefore, the obtained breakage fraction can be taken as equal to the breakage probability. The method implies, however, that the initial sieve fraction is narrow enough, so that all fragments of a broken particle are passing the lower sieve mesh.

Determination of breakage probabilities from laser diffraction data has two advantages: first, the required amount of substance and thus also the testing time can be decreased noticeably compared to sieve analysis. Second, the determination of the size distribution by sieving will become increasingly inefficient if not impossible for smaller particles sizes due to the more cohesive behaviour of fine powders. However, the evaluation of laser diffraction data is more complicated. Here, already the initial sample possesses a size distribution. After impacting, this distribution has shifted and has become broader, and the fraction of broken particles is not as obviously determined as in the sieve analysis. Therefore, a population balance model was employed. Here, the breakage function is coupled with the breakage probability to obtain the size distribution of the powder after milling. For each size class, the mass change during comminution is given by

$$\Delta m_i = \sum_{j=1}^{i-1} S_j \cdot m_j b_{ij} - S_i \cdot m_i \qquad (3.5)$$

where i and j denote different size classes; m_i, m_j are the masses of particles in class i and j respectively, S_i is the breakage probability for particles of class i, b_{ij} is the breakage coefficient from class j to i (Vogel and Peukert, 2002). Since a narrow size distribution was used as feed material, the breakage probability S_i is assumed to be constant for all particle classes at one impact velocity (this assumption is implicitly also made in the sieve analysis method). To obtain the breakage coefficients, a breakage function B(y,x) has to be modeled, which describes the particle size distribution of a size class after breakage. For this model, a distribution given by Broadbent and Callcott (1956) was chosen:

$$B(x,y) = \frac{1 - e^{\left(-\frac{x}{y}\right)}}{1 - e^{-1}}, \quad 0 < x \le y \qquad (3.6)$$

Here, x is the particle size after breakage, y the initial particle size. The advantage of the Broadbent-Callcott equation is that it contains only one parameter, the initial particle size y. Using this breakage function, a size distribution $q_B(x)$ of a completely comminuted sample can be calculated from the distribution of the feed material, $q_F(x)$. The size distribution $q_P(x)$ of a partly broken sample can then be calculated by

$$q_P(x) = S \cdot q_B(x) + (1 - S) \cdot q_F(x) \qquad (3.7)$$

The thus obtained size distribution $q_p(x)$ is fitted to the experimentally obtained distribution by the least-square method, with S being the variable parameter. The interval sizes, as they are obtained from the measurement device (Malvern Mastersizer, Malvern, UK), increase linearly with particle size. Thus, if a non-weighted fitting of the obtained $q(x)$ data is performed, the weighing of the fine particles would be too high, while, in fact, the most important part of the distribution curves for determination of the breakage probability is the region of the "non-milled" particle sizes, i. e. the coarse particles. Therefore, a weighing of the intervals needs to be introduced by multiplying $q(x)$ with the interval size Δx, and the actual curve fitting was performed with the weighed distributions $q(x)\Delta x$. Note that this quantity $q(x)\Delta x$ represents the percentage of volume (or mass) that lies within a given interval and is an appropriate basis for the evaluation. An example for the curve fitting strategy is shown in Fig. 3.5. The empirical correlation coefficient R^2 was used to judge the fit quality. The values for R^2 were found to be around 0.9 or higher, which was considered acceptable. For a quick evaluation of experimental data, the described model was implemented into a Microsoft Excel sheet. The fitting itself was done using the solver function of Excel.

A more exact approach to the evaluation of breakage probability would be to directly count the individual particles, i. e. to measure and to evaluate the number distribution of the powders. Suitable methods for this could be e. g. image analysis, scattered-light particle sizing or the electrical sensing zone method (also known as Coulter principle), all of them having disadvantages as far as the size range or the experimental effort are concerned. Therefore, due to the availablity of the laser diffraction and its ease of operation it was chosen to use the method described above.

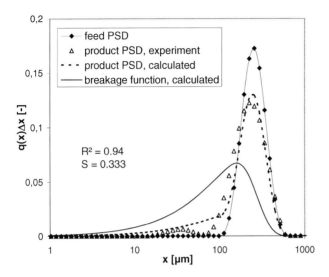

Fig. 3.5 Example of measured and calculated weighted particle size distributions (α–lactose monohydrate, impacted at 56 m/s); note: strictly speaking, it is not correct to display a q(x)Δx distribution as a continuous line - it is done here for reasons of legibility; q(x)Δx instead of q(x) is shown in order to better visualise the fitting process

The dispersion of particles in the nozzle bears the danger that some particles are already comminuted during the dispersion. Thus, the "apparent" breakage probability, S_{app}, which is measured after the experiment, contains the comminution both on the impact plate and in the nozzle. Therefore, the comminution in the nozzle was determined separately by additional experiments where the impact plate was taken out and the particles were only fed through the dispersion nozzle, accelerated along the tube and collected in the filter. Then, the mass fraction of particles that were comminuted in the nozzle could be determined. These experiments were repeated for different particle velocities, and an empirical equation was used to describe the breakage in the nozzle in dependency on the final impact energy. Finally, the "true" breakage probability of a particle at one single impact on the impact plate can be determined using the following equation:

$$S_{plate} = \frac{S_{app} - S_{nozzle}(W_{kin})}{1 - S_{nozzle}(W_{kin})} \qquad (3.8)$$

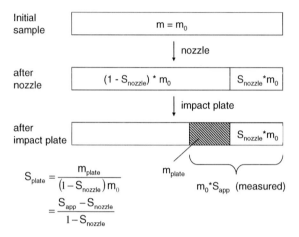

Fig. 3.6 Derivation of Eq. (3.8)

Eq. (3.8) can be easily derived with the help of the sketch in Fig. 3.6: S_{nozzle} is the fraction of particles that are comminuted in the nozzle. S_{app}, the "apparent breakage probability", is the entire fraction of broken particles that is measured after the experiment, i.e. particles that were comminuted in the nozzle and particles comminuted on the impact plate. S_{plate} represents the particles that reach the impact plate unbroken, and are comminuted by impacting the plate. The initial mass for this impact has to be corrected by the factor $(1-S_{nozzle})$ in order to exclude those particles that were already broken before the impact. Fig. 3.7 illustrates the correction process for the case of the easily-breaking acetylsalicylic acid. Here, the initial comminution is already relatively high, and therefore the correction has a noticeable influence on the final result. It should be mentioned that this extreme example of acetylsalicylic acid was obtained with the old nozzle design, where a bended piece of tubing was present after the nozzle. After the nozzle design was improved, the initial comminution was strongly decreased: even at high pressures, comminution in the nozzle was below 13 % for acetylsalicylic acid. (See Appendix A for detailed comminution data of ASA with the old and new nozzle design.)

Compared to sieve analysis, the laser diffraction method should be expected to be more sensitive to chipping: because the size resolution of the measured distribution is better, this method should be more sensitive to only small changes in particle size as happens when only slight chipping takes place. Therefore, slightly higher breakage probabilities should be obtained as compared to sieve analysis. In fact, the comparison of both method shows only slight deviations of the results from both methods, as will be explained in Chapter 4.2.4.

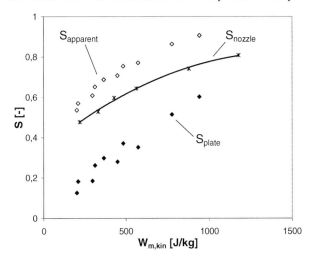

Fig. 3.7 Correction of the measured breakage probability for acetylsalicylic acid (sieve data); an empirical fit equation can be given for comminution in the nozzle: $S_{nozzle} = -2.15 \cdot 10^{-7} \ kg^2/J^2 \cdot W_{m,kin}^2 + 6.43 \cdot 10^{-4} \ kg/J \cdot W_{m,kin} + 0.35$ ($R^2 = 0.998$)

3.2. Nanoindentation

3.2.1. Device and measurement conditions

Indentation experiments were done with a Nanoindenter XP by MTS Systems, Oak Ridge, Tennessee, USA. This device applies the load by electromagnetic actuation: an electric current is passing through a coil that sits in an annular magnet; the force is imposed on the indenter shaft that is located inside this magnet. The displacement is sensed with a three-plate capacitance gauge. The producer gives resolutions of 50 nN for the force and < 0.01 nm for the displacement.

A Berkovich indenter was used for the experiments, which is the most commonly used tip geometry nowadays. It has the shape of a three-sided pyramid with a half-angle of 65.3°. Its main advantage compared to a (four-sided) Vickers indenter is the easier manufacturing of a sharp tip; therefore, more exact results can be obtained especially at lower loads, as was shown by Dukino and Swain (1992). The calibration constants that are given in the model equations for fracture toughness as given by Ponton and Rawlings (1989a, 1989b) are valid for a Vickers indenter. For a Berkovich indenter, the constant needs to be adapted. Dukino and Swain obtained a modifying factor by applying the ratio of normalised stress intensity factors $k_1(n)$ after Ouchterlony (1976) for the two indenter geometries:

$$x_b = \frac{k_1(4)}{k_1(3)} x_v = 1.073 x_v \qquad (3.9)$$

with

$$k_1(n) = \frac{\sqrt{\dfrac{n}{2}}}{\sqrt{1 + \dfrac{n}{2\pi} sin(\dfrac{2\pi}{n})}} \qquad \text{for} \quad n \geq 2 \qquad (3.10)$$

Here, x_b is the calibration constant of the Berkovich indenter, x_v the calibration constant for a Vickers indenter, n is the number of cracks generated by the indenter.

For the measurements in this work, the nanoindenter was operated in the so-called "continuous stiffness mode" (CSM™). In this mode, the tip is pressed into the material according to a predefined loading profile, while a harmonic oscillation is superimposed onto the tip movement. This leads to a repeated unloading during indentation, from which Young's modulus and hardness can be determined using Eqs. (2.21) to (2.23). Thus it is possible to obtain Young's modulus and hardness at all indentation depths during one single experiment. Loads of up to 700 mN were applied in this work. For the determination of hardness and Young's modulus, only indentation depths of around 2000 nm (corresponding to a few 10 mN, depending on the material) are necessary; however, for the determination of fracture toughness, cracks need to be induced in the material, and higher loads were required for most of the substances. For evaluation of the crack length, pictures were taken after the indentation with a light microscope (Leitz Ergolux, Leitz GmbH, Wetzlar, Germany), and the indent diagonal a and the crack length l were measured from these pictures.

3.2.2. Sample preparation

Nanoindentation measurements have some requirements with regard to the substrate, i. e. the particle that has to be measured. The so-called 10 %-rule states that the indenter tip should not move more than 10 % of the substrate thickness into the material, otherwise the result can be influenced by the properties of the underlying material. Since it was found that indentation depths up to 2000 nm are enough for the determination of hardness and Young's modulus, a particle size of 20 µm would be sufficient; for fracture toughness determinations, larger particles are needed, because usually deeper indents are needed to induce cracks. For a proper evaluation of the crack length, the cracks should not split the entire crystal. Experience showed that a particle size of at least 100 µm is desirable. Generally, larger particles facilitate the handling during preparation.

The particle needs to have a flat surface, so that the Oliver-Pharr model can be applied properly. Further, also the tilt of the surface can affect the measurement result, therefore the surface should be as flat as possible, a slope of less than 3 % should be attained. To eliminate this source of error, all particles were checked under an optical microscope after preparation. The particles were considered acceptable for indentation, if at magnifications between 25-50x the visible area under the light microscope could be brought into focus at the same time (with higher slopes, only a small band will be visible sharply at higher magnifications, which moves when the microscope table is moved up and down).

Further, the particle needs to have a proper support on the sample holder: it should not move or roll away, and the material that carries the particle should be more rigid than the particle itself. To achieve this goal, various methods were employed to fix the particles to a supporting glass slide. These different methods are explained in detail in Appendix A, along with their various advantages and disadvantages. The choice of the method depends on the material that has to be prepared. Where possible, particles were used as delivered (acetylsalicylic acid, citric acid, sucrose, ascorbic acid). Because smooth surfaces are required, some substances had to be recrystallised. Recrystallisation results not only in smoother surfaces, but also in regular crystal shapes, which in turn simplifies handling and glueing of the particles. Also, larger particle sizes can be achieved by recrystallisation. As a general rule, the particles should be crystallised as large as possible, since the handling becomes much more easy.

3.3. Tabletting

Tabletting experiments were performed using a Zwick Z020/TN2S universal testing machine (Zwick GmbH, Ulm, Germany) with a circular punch having a diameter of 9 mm. The software controlling the machine registers the applied force and the displacement of the punch during a compression experiment. Unfortunately, data can not be recorded on removal of the punch, as is done usually in special tabletting machines. Experiments were done with the initally selected four model substances, α-lactose monohydrate, acetylsalicylic acid, and Compounds A and B. Different particle sizes were studied with each material, both "in-die" and "out-of-die" experiments were done. Pressures up to 240 MPa were applied, the punch velocity was 50 mm/min. The amount of powder used for each tablet was chosen in a way that a tablet height of 3 mm would be obtained if the tablet had a porosity of zero. For the evaluation of "out-of-die" experiments, the tablet height was measured with a micrometer screw one day after compression. For "in-die" experiments, the tablet height was calculated from the punch displacement data.

3.4. Shear testing

The shear tests in this work were done with a Schulze Ring Shear Tester RST-01.01 (see Fig. 3.8) from Dr.-Ing Dietmar Schulze Schüttgutmesstechnik, Wolfenbüttel, Germany. In this device, an annulus is filled with a sample of powder and covered with a lid, and a normal force is applied. The preconsolidation of the bed is done by rotating the lid against the annulus until a steady state flow is achieved. The annulus and the lid have a specially designed waffle surface to make sure that the shear plane lies within the powder bed, and not at the interface between powder and apparatus. After preconsolidation, the powder bed is sheared again under different normal stresses.

Measurements were done with a shear cell type SV10 (sample volume approx. 100 ml). Measured were the flowability of the powders as well as the wall friction on two steel wall samples (steel type 1.4404) with different surface roughness and on polytetrafluorethane (PTFE). The standard testing procedures were used as they are described in the user manual of the device:

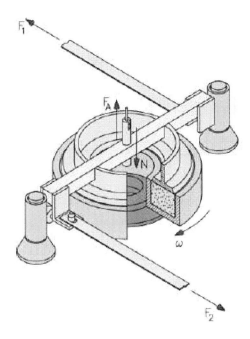

Fig. 3.8 Schulze Ring Shear Tester (picture from the manufacturer)

For the determination of flowability, the yield locus was determined for three different consolidation stresses, at normal stresses of 2500, 5000 and 10000 Pa. For each consolidation stress, three points of the yield locus were determined, with applied normal stresses of 20 %, 50 % and 80 % of the normal consolidation stress (in that order). These measurements for one yield locus are done with the same powder bed, which is consolidated again before measuring the next point.

For the wall friction measurements, a maximum normal load of 10000 Pa was used. Measurement points at six different loads were obtained. The shear stress was first measured at maximum load, then the load was decreased in steps of 1800 Pa down to 1000 Pa and subsequently increased to the maximum again. All samples were measured twice in order to ensure reproducibility.

3.5. Atomic Force Microscopy

The principle of an Atomic Force Microscope is shown in Fig. 3.9. A cantilever is located above a substrate. The substrate is moved up and down to and from the cantilever by piezo actuators. The deflection of the cantilever is monitored by the deviation of a laser beam that is reflected from the backside of the cantilever to a photodiode.

For the determination of particle adhesion forces, a single particle is glued to the tip of a cantilever. A typical force-distance curve of such a measurement is depicted in Fig. 3.10. The substrate is approaching the AFM cantilever with the attached particle. If the distance between particle and substrate becomes small enough, the attractive forces become higher than the spring force of the cantilever; the cantilever bends down and the so-called "jump-in" takes place. If the substrate is raised further, the particle is pressed onto the substrate and the cantilever is bent upwards. If the substrate is retracted, the particle stays in contact with the substrate, the cantilever is bent down and exerts a tractive force onto the particle-substrate connection. If this force exceeds the adhesive force, the particle is separated from the substrate, the so-called "jump-out" is happening. The adhesion force is determined from the spring force of the cantilever at the jump-out, which can be calculated after Hooke's law by multiplying the cantilever's spring constant with the distance between jump-out and zero-force-point.

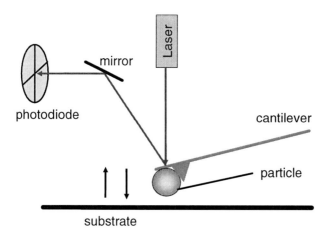

Fig. 3.9 Principle of AFM

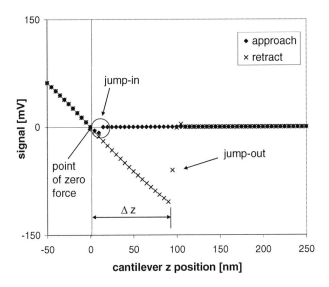

Fig. 3.10 Typical AFM force-distance curve

The spring constant of the cantilever can be determined by the "added mass" method: the resonance frequency of the cantilever with an adhering mass is determined. Glass spheres between 10 and 30 μm diameter were used as added mass. From the resonance frequencies, the spring constant k can be determined:

$$k = \frac{4\pi^2 M}{\left(\frac{1}{f_1^2} - \frac{1}{f_0^2}\right)} \qquad (3.11)$$

where M is the added mass, f_0 is the resonance frequency of the cantilever alone, f_1 is the resonance frequency of the cantilever with added mass.

The particles are glued to the cantilever under a light microscope (Leitz Ergolux, Leitz GmbH, Wetzlar, Germany), with the use of a micro-manipulator. A two component epoxy glue was used (UHU endfest 300 plus, UHU GmbH & Co. KG, Bühl, Germany). For acetylsalicylic acid, which is easily soluble in almost any solvent, a faster binding glue was used (UHU plus schnellfest).

The measurements were done with a commercial AFM, Nanoscope IIIa Multimode by Veeco Instruments Inc., Plainview, NY, USA. Scans of the surfaces were taken in Tapping Mode (TM), adhesion force measurements were done in Contact Mode (CM). Force distributions were measured using the "auto ramp" function of the controller software, on a lattice of 15 x 15 points, spaced 1 μm. The measurement cell was flushed with dry nitrogen. For experiments at higher levels of relative humidity, the nitrogen stream was mixed with a humidified nitrogen stream to achieve the desired humidity in the cell.

3.6. Inverse Gas Chromatography

For the iGC experiments, a commercial gas chromatograph (PerkinElmer Auto System GC, PerkinElmer Inc., Waltham, MA, USA) was modified to obtain the following iGC setup which is illustrated in Fig. 3.11: Helium is used as carrier gas; a part of the gas flows through a bubble flask that is filled with the probe gas and held at a constant temperature. Thus the gas is approximately saturated. A portion of this saturated stream is then injected into the helium carrier stream with an automated 6-port valve (sample loop size = 0.1 ml). The carrier gas flows through the column containing the solid, which is located in the oven. The columns used were stainless steel HPLC columns with an inner diameter of 4 mm and a length of 125 or 250 mm, delivered from Knauer GmbH, Berlin, Germany. After the column outlet, a flame ionisation detector (FID) detects organic compounds in the gas. The data acquisition and the 6-port valve are controlled by the PerkinElmer software TotalChrome 6.2. Each column was packed very carefully, the powder was filled into the column spoon by spoon and regularly tapped onto the work bench in order to densify the bed evenly and to avoid cavities in the packing that later may lead to gas holdup during the experiments. For each sample, at least two columns were packed and tested, to make sure that the packings were correct and the results reproducible. As probe gases, different unbranched alkanes (n-hexane, n-heptane, n-octane and n-nonane) as well as ethanol and acetone were used. The dead time was determined using methane. Flow rates of the carrier gas were set between 5 to 10 ml/min. Measurements were done at 30°C, some additional experiments were done at 50°C.

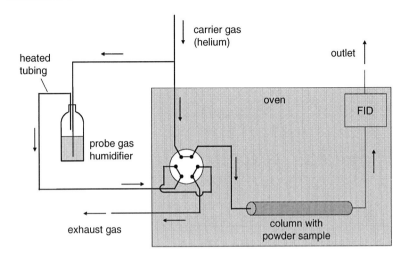

Fig. 3.11 Experimental setup for the inverse gas chromatography

3.7. Materials

Four pharmaceutical substances were initially chosen for the experiments. These substances are α-lactose monohydrate and acetylsalicylic acid ("ASA"), that were chosen as relatively harmless (non-toxic) model substances, that are already well-known in the literature. Two active pharmaceutical ingredients were obtained from Novartis Pharma AG, Basle, Switzerland. These substances were chosen because they exhibit a critical behaviour during milling trials: Compound A is a substance to be used in dry powder inhalation formulations, therefore particles in the micrometer size range are desired. However, even at high energy inputs in a spiral jet mill, it is very hard to achieve comminution of that substance. Compound B exhibits a very sticky behaviour: in milling trials, the mill gets clogged by Compound B after very short times, and a steady process can not be kept running.

Later in this work, a broader data basis was desired for comminution experiments and nanoindentation, therefore more organic substances were chosen for those methods. All substances chosen are crystalline, consisting of small organic molecules with molecular weights between 75 and 810 g/mol. Some experiments were also done with other types of materials, namely polymers and inorganic substances. Table 3-1 gives an overview over all substances studied and their initial particle sizes.

Table 3-1 Overview over all studied materials

Substance	Size range [*] (sieve fraction)	x_{50} [**] [µm]	Supplier [***]
Organic crystals			
α-lactose monohydrate	from 45-63 µm	72	1; 2
	to 160-200 µm	251	
Acetylsalicylic acid	355-500 µm	506	3
Compound A	125-500 µm	288	4
Compound B	n.a.[*]	85	4
L(+)-Ascorbic acid	250-355 µm	381	5
Sucrose	355-500 µm	548	6
L(+)-Tartaric acid	250-355 µm	367	5
Glycine	250-355 µm	355	5
Polymers			
PS 144C	2.5-2.8 mm [****]	-	7
PS168N	2.5-2.8 mm [****]	-	7
Inorganic materials			
Limestone	355-500 µm	504	8
Ammonium sulphate	500-710 µm	750	5
Potash alum	from 0.7-1.0 mm	-	5
	to 2.8-3.2 mm [****]	-	
Compound materials			
Epoxy powder coating	355-500 µm	508	9
Polyester powder coating	1.8-2.0 mm [****]	-	9

* Size range as used as feed for the single particle comminution experiments; with Compound B, sieving was not possible due to the rod-shape of the particles

** determined from laser diffraction

*** Suppliers: 1: DMV International, Veghel, Netherlands; 2: Meggle GmbH & Co. KG, Wasserburg, Germany; 3: Rütgers Chemicals, Castrop-Rauxel, Germany; 4: Novartis Pharma AG, Basel, Switzerland; 5: C. Roth GmbH, Karlsruhe, Germany; 6: Südzucker AG, Mannheim, Germany; 7: BASF AG, Ludwigshafen, Germany; 8: Endress GmbH & Co. KG, Gräfenberg, Germany; 9: Herberts Pulverlack GmbH, Essenbach, Germany

**** Data taken from comminution experiments by Vogel and Peukert (2003)

4. Results and Discussion

4.1. General characterisation of the powders

4.1.1. Macroscopic milling behaviour

With the four initially chosen substances, preliminary milling tests have been done at the milling station of Novartis Pharma AG in Basle, Switzerland. Two types of mills were used: a Condux CST-10 pin disk mill (rotor diameter = 76 mm) and an Escolab JRMS 80 spiral jet mill. Three levels of power input were applied with each mill: 8000, 13500 and 19000 rpm with the Condux mill (corresponding to circumferential rotor speeds of 32, 54 and 76 m/s), and nitrogen pressures of 2, 4 and 6 bar with the Escolab mill. Generally, the spiral jet mill supplies more energy to the product than the pin mill; also, the following results show that a higher extent of comminution is achieved with the spiral jet mill already at 2 bar than with the pin mill at 19000 rpm. More detailed considerations about the specific power input in the mills, along with quantitative estimates can be found in Appendix C.

Several 100 grams of substance were milled in each experiment. For Compounds A and B, experiments were only done at 19000 rpm and 2 and 6 bar. The feed rate was kept between 0.6 and 1 kg/h. The material was fed by hand, therefore variations in the feed rate occurred. After milling, the particle size distributions were measured by laser diffraction (Sympatec Helos H067, Sympatec GmbH, Clausthal, Germany), and the BET surface was determined (Nova 2200, Quantachrome Corp., Boynton Beach, FL, USA). Two different batches of lactose have been used because the first one (DMV S00200) turned out to possess a very broad size distribution, which made it more difficult to compare the results with other substances. Therefore, lactose of the specification SV001 (also from DMV) was used for further experiments. This specification as well as the other feed materials possess a narrow monomodal size distribution.

Table 4-1 Overview over initial milling tests

a) Lactose S00200

milling conditions	x_{50} [μm]	$x_{50}/x_{50,0}$ [-]	S_{BET} [m²/g]	$S_{BET}/S_{BET,0}$ [-]
before milling	50.2	-	0.39	-
8000 rpm	51.7	1.0	0.45	1.2
13500 rpm	44.7	1.1	0.47	1.2
19000 rpm	34.9	1.4	0.61	1.6
2 bar	15.9	3.2	1.26	3.2
4 bar	11.8	4.3	1.55	4.0
6 bar	8.1	6.2	1.81	4.6

b) Lactose SV001

milling conditions	x_{50} [μm]	$x_{50}/x_{50,0}$ [-]	S_{BET} [m²/g]	$S_{BET}/S_{BET,0}$ [-]
before milling	238.7	-	-	-
8000 rpm	203.1	1.2	-	-
13500 rpm	151.6	1.6	-	-
19000 rpm	63.5	3.8	-	-
2 bar	109.9	2.2	-	-
4 bar	15.0	16.0	-	-
6 bar	10.7	22.3	-	-

c) Acetylsalicylic acid

milling conditions	x_{50} [μm]	$x_{50}/x_{50,0}$ [-]	S_{BET} [m²/g]	$S_{BET}/S_{BET,0}$ [-]
before milling	502.2	-	0.08	-
8000 rpm	127.8	3.9	0.21	2.6
13500 rpm	32.2	15.6	0.38	4.8
19000 rpm	18.0	27.9	0.51	6.4
2 bar	5.9	85.3	0.97	12.1
4 bar	4.3	118.2	1.51	18.9
6 bar	3.2	156.5	0.86	10.8

Table 4-1 (continued)

d) Compound A

milling conditions	x_{50} [μm]	$x_{50}/$ $x_{50,0}$ [-]	S_{BET} [m²/g]	$S_{BET}/$ $S_{BET,0}$ [-]
before milling	417.4	-	0.11	-
19000 rpm	209.5	2.0	0.38	3.5
2 bar	301.3	1.4	0.25	2.3
6 bar	254.6	1.6	0.70	6.4

e) Compound B

milling conditions	x_{50} [μm]	$x_{50}/$ $x_{50,0}$ [-]	S_{BET} [m²/g]	$S_{BET}/$ $S_{BET,0}$ [-]
before milling	30.7	-	0.64	-
19000 rpm	19.5	1.6	0.77	1.2
2 bar	18.6	1.6	0.85	1.3
6 bar	14.7	2.1	1.10	1.7

The resulting particle size distributions are depicted in Figs. 4.1 - 4.5. The median values x_{50} of the distributions as well as the BET surface and the change of these quantities compared to the feed material are given in Table 4-1. As can be expected, particle sizes are decreasing with increasing milling intensity. Comparing the results for the two different lactose batches (Fig. 4.1 and Fig. 4.2), a difference between comminution at 2 bar in the spiral jet mill and at 19000 rpm in the pin mill is observed: while for batch S0020, the spiral jet mill achieves a finer product, the pin mill gives a finer product for batch SV001. This may be a consequence of the different particle size distributions of the feed material, but also variations in the feed rate may be a likely reason: feeding was done manually, thus there were some fluctuations in the loading of the mill and in the specific energy input that may also lead to different comminution results.

Despite some uncertainties concerning the specific energy input, major differences between the different substances can be seen: Acetylsalicylic acid (Fig. 4.3) was comminuted very easily; compared to lactose, far less power was necessary to achieve a similar size reduction ratio and increase in BET surface area. After milling, a noticeable electrostatic charging was observed for ASA.

For Compound A, there was not much substance available. Therefore, only portions of 50 g were milled. Compound A did not stick to the walls, but was hardly reduced in size even at high power input, as the size distributions show (see Fig. 4.4). In fact, comminution happened to some extent already at very low energy inputs: Fig. 4.6. shows particles of Compound A as they were received from crystallisation: several large particles are aggregated and grown together in one point. These aggregates are broken very easily - applying ultrasound during the PSD measurements is sufficient (compare to Fig. 4.4). But once these aggregates are destroyed, further comminution of Compound A is hardly achieved.

Compound B was found to be very prone to scaling at the walls of the mill. The milling trials had to be stopped after a short time because the mill was clogged: the milling chamber was filled with compressed substance sticking to the walls. Also here, electrostatic charging was observed. The powder was only slightly reduced in size, as can be seen from Fig. 4.5.

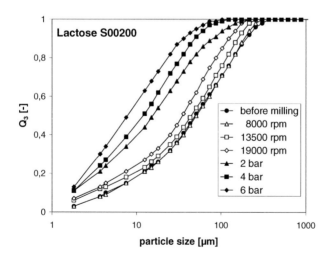

Fig. 4.1 Particle size distributions of lactose S00200 before and after milling

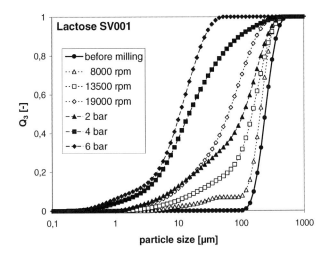

Fig. 4.2 Particle size distributions of lactose SV001 before and after milling

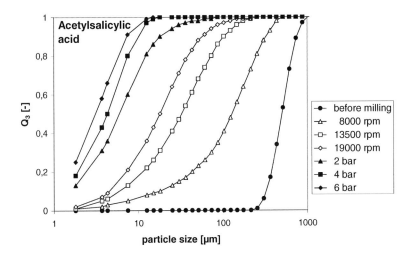

Fig. 4.3 Particle size distributions of acetylsalicylic acid before and after milling

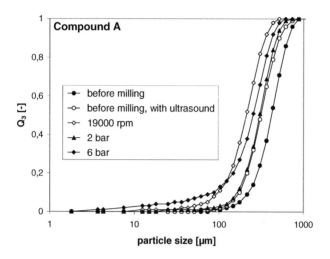

Fig. 4.4 Particle size distributions of Compound A before and after milling

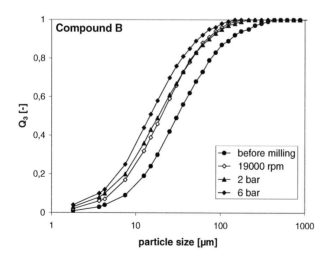

Fig. 4.5 Particle size distributions of Compound B before and after milling

Fig. 4.6 Compound A before milling (left) and after milling at low energy input (right)

These preliminary tests illustrate the macroscopic milling performance of the materials: ASA can be comminuted very easily, lactose exhibits a medium performance. Compounds A and B can be hardly comminuted; in the case of Compound A, this seems to be attributed to an unfavorable breakage behaviour of the material, while for Compound B, the poor milling performance might be due to the transport properties that prohibit an efficient milling process. These observations will be confirmed and quantified by single particle comminution experiments in the following chapters.

4.1.2. SEM

SEM images of the powders were taken before milling and after milling in a spiral jet mill. The images are shown in Fig. 4.7 to Fig. 4.10. Acetylsalicylic acid consists of large, plate-shaped crystals before milling; after milling, the size is noticeably reduced, the particle shape remains similar. In the cases of lactose and Compound A, the particles possess an irregular shape already before milling. The number of fines adhering to the surfaces is increased after milling. Compound B consists of very smooth rod-shaped crystals; breakage during milling seems to happen easily perpendicular to the long axis.

Fig. 4.7 Acetylsalicylic acid before (left) and after milling in a spiral jet mill at 6 bar pressure (right)

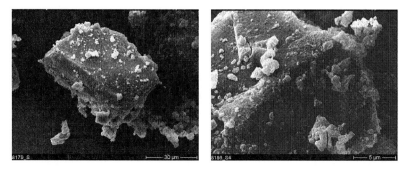

Fig. 4.8 α-lactose monohydrate before (left) and after milling in a spiral jet mill at 6 bar pressure (right)

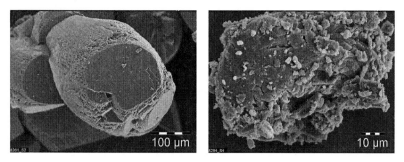

Fig. 4.9 Compound A before (left) and after milling in a spiral jet mill at 6 bar pressure (right)

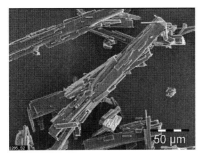
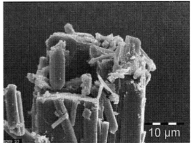

Fig. 4.10 Compound B before (left) and after milling in a spiral jet mill at 6 bar pressure (right)

4.1.3. Shape analysis

Shape analysis was done using a light microscope and the image analysis software analySIS 5.0 (Soft Imaging Systems GmbH, Münster, Germany). Aspect ratios of the powders before and after milling were determined. The mean aspect ratios are shown in Fig. 4.11. The results confirm what can be seen from the images: Non-milled Compound B possesses by far the highest aspect ratio (2.4 compared to 1.7-1.8 for the other substances) due to its rod-like shape. On milling, the aspect ratio is strongly reduced for Compound B, but only to a noticeably lesser extent for the other substances; still, Compound B remains the powder with the highest aspect ratio of the four substances.

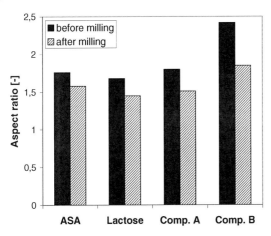

Fig. 4.11 Mean aspect ratios of the powders before and after milling

4.2. Single particle impact comminution

4.2.1. Characterisation of the air jet device

First tests with the new air jet device were performed using glass spheres. Fig. 4.12a shows the measured velocity distributions of glass spheres with a mean diameter of 90μm, impacting onto the plate at 90°. Clearly, multiple peaks can be observed, a consequence of the air flowing against the plate. After the first impact at about 50 m/s, the particles bounce back from the impact plate and cross the measurement volume again, this time with a negative sign due to the changed direction of flight. Because the air stream now decelerates the particles, the absolute velocity slightly decreases (see Fig. 4.12a - about -45 m/s in this example). Further on, the particle is accelerated again towards the impact plate and crosses the measurement volume another time, this time with a noticeably slower speed. The peak in the velocity distribution at around -20 m/s represents the second back-bouncing of the particle.

From this measurement it becomes clear that with this setup, no defined stressing conditions can be achieved. It has to be made sure that particles can leave the impact zone after the first impact without hitting the plate again. Therefore, the impact plate was tilted, so the particles can leave the impact zone after impact by flying sideways. The tilt has to be sufficient so that the horizontal component of the particle velocity is high enough to let the particles leave the region of the air jet. From measurements at different angles, it was found that at angles smaller than 75°, no more secondary positive peaks (meaning another flight path towards the plate) were detected. To be absolutely sure, an impact angle of 60° was chosen as standard for the experiments. A velocity profile, measured with glass spheres at this angle, is shown in Fig. 4.12b. Now, only a very small peak with negative velocity components is found, which comes from particles rebouncing from the plate but flying to the side.

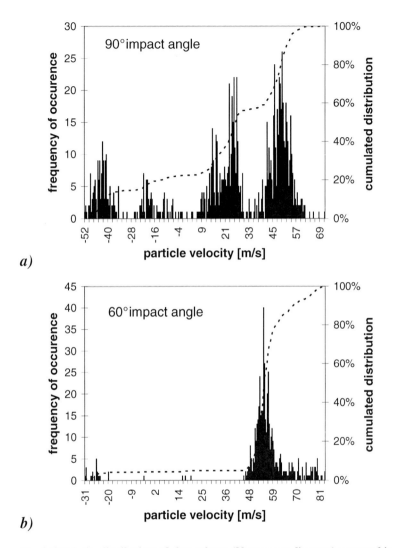

a)

b)

Fig. 4.12 Velocity distributions of glass spheres (90 μm mean diameter) measured in the impact jet mill in vertical direction above the impact plate: a) impact angle = 90°, multiple impacts are detected; b) impact angle = 60°, only single impacts take place

A typical velocity distribution for an experiment with α-lactose monohydrate particles (160-200 μm) is shown in Fig. 4.13. Again, the negative velocity values stem from particles that fly sideways through the measurement volume after the impact. The positive values are divided into two distinct peaks. The higher velocities correspond to fine particles that were already present in the feed, adhering to the surfaces of the coarse material. These fines were observed in several of the studied substances. The fines are accelerated in the air stream much faster than the coarse material and reach almost the velocity of the air. The lower velocity represents the "true" particle velocity, i.e. the velocity of the coarse particles.

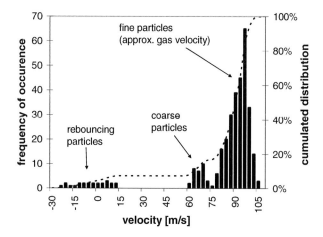

Fig. 4.13 Velocity distributions of α-lactose monohydrate, impact plate set to 60°

From inspection of the distribution curve it seems as if many more fines are present in the sample than coarse particles. But one has to keep in mind that with LDA, the signal quality decreases for bigger particles; irregularly shaped particles of several 100 μm diameter often give signals that can not be properly evaluated from the software and therefore are disregarded. Therefore, relatively few fine particles can result in more measured signals than many coarse particles. But nevertheless, these fine particles can be neglected in terms of weight fraction in the evaluation of the experiments. For further evaluations, the impact velocity of the "coarse" particles was used. The finer the feed material gets, the more are the velocities of coarse and fine particles overlapping, and the determination of the velocity of the coarse material becomes problematic. Therefore it will be very helpful to find a way to determine the particle velocity from the initially applied air pressure. Thus, the elaborate LDA measurements and evaluations could be spared. A detailed characterisation of particle and gas velocities in dependence of the applied air pressure can be found in Appendix D.

4.2.2. CFX modelling

For a proper determination of the impact energy of the particles, both the impact velocity and the impact angle have to be known. Due to the geometry of the setup, the velocity can be measured only a few millimeters above the impact plate (1.2 – 3 mm). Because of the stagnation point flow field near the impact plate, particles will be slowed down by the air stream when approaching the impact plate, and will be deflected from their straight path. Therefore, the true impact velocity will be lower than the measured value, and the true impact angle will deviate from the nominal angle to which the impact plate is set. In order to check whether these effects are negligible for a certain particle size, preliminary trajectory calculations were done with a self-written C++ programme for a 90° impact. Here, an axisymmetric stagnation point flow was assumed, where an analytical solution for the flow field exists (Schlichting, 1987). For a more detailed analysis, especially for feed particle sizes below 100 μm, CFD calculations were done for particle impacts at 60°. For these calculations, ANSYS CFX 10.0 was used, employing the shear stress model and one-way coupled fluid-particle interactions. Particles were assumed to start with 0.1 m/s at the beginning of a 400 mm acceleration tube. Particle density, particle sizes and the gas velocity were varied.

For an estimate of the velocity distribution of the particles dispersed over the tube diameter, impact velocities were determined for 180 particle tracks, distributed randomly over the diameter. The obtained velocity distributions were found to possess standard deviations between 3.6 % (100 μm particles) and 4.5 % (30 μm particles), smaller than the measured velocity distributions that amount to about 10 % (see Chapter 4.2.6). The worst case, i. e. the strongest deflection of a particle track, happens if a particle leaves the tube at the outer diameter. In this case, the deviation from the set impact angle is the largest, and the decrease of the vertical velocity during the last 2 mm above the impact plate is the strongest. These worst-case deviations of impact angle and impact speed are shown in Fig. 4.14, where they are plotted over the dimensionless Stokes parameter Ψ. This parameter is defined as

$$\Psi = \frac{\rho_p x^2 v_P}{18 \eta D},$$
(4.1)

where ρ_p is the particle density, x is the particle size, v_P the velocity of the incoming particle, η is the dynamic viscosity of the fluid and D is a characteristic length of the system, in this case the distance between the pipe outlet and the impact plate (D = 4.7 mm).

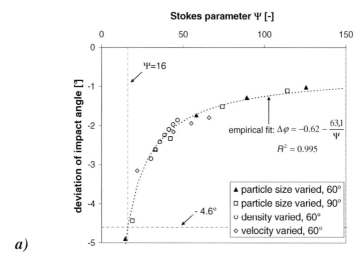

a)

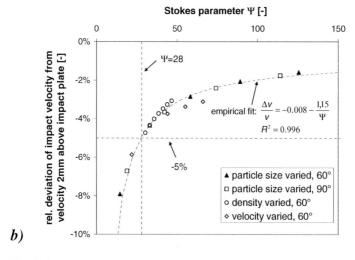

b)

Fig. 4.14 a) Worst case deviations of impact angles; b) worst case deviations of particle velocities; (data for 60° impact angle: from CFX modelling; 90° impact angle: from preliminary calculations using a simplified flow field)

From Fig. 4.14 it can be seen that, when varying different parameters, the worst case deviations are very similar for the same Stokes parameter. For the worst-case deviations, it was decided to accept deviations of impact speed up to 5 %, which corresponds to a deviation of impact angle of up to 4.6° in the case of an impact plate set to 60°. The reduction of particle velocity is below 5 % for Stokes parameters bigger than 28. The maximum allowed deflection of the impact angle is reached with a Stokes parameter even as low as 16. This leads to the conclusion that reasonable results -with an acceptable standard deviation- can be obtained with the impact device for $\Psi > 28$. This value corresponds, for example, to a 25 µm lactose particle ($\rho = 1540$ kg/m³) leaving the acceleration tube with 45 m/s. According to this, it will be possible to characterise powders down to a few 10 microns, the exact size depending on the density of the material and the minimum impact speed required to break the particles. The feasible size range could even be shifted further towards smaller sizes by installing a vacuum pump, i. e. operating the milling chamber at lower pressure (< 0.1 bar) in order to reduce the drag force. All measurements shown in this work were done with particles of similar density as lactose, and particle sizes of 45 µm and higher, so the Stokes parameters of all measurements were significantly higher than 28. Deviations due to the stagnation point flow can therefore be neglected.

4.2.3. Comparison of the air jet device with the Schönert device

The obtained breakage probabilities were evaluated using the approach by Vogel and Peukert (2003): several experiments were performed at different impact speeds. Thus, a dependency of S over $W_{m,kin}$ is found, and a curve fit using Eq. (2.2) yields the two parameters f_{Mat} and $xW_{m,min}$ that describe the breakage behaviour of the material. The specific kinetic energy is determined from the impact velocity by the equation

$$W_{m,kin} = \frac{1}{2} v^2 \sin^2 \varphi \tag{4.2}$$

where φ is the impact angle. This means that for oblique impacts, only the velocity component normal to the impact plate is considered. In the following, the modified parameters $W^*_{m,kin}$, f^*_{Mat} and $xW^*_{m,min}$ that account for the elasticity of the impact plate according to Eqs. (2.10), (2.13) and (2.14) will be used.

For the comparison of the two different single particle devices, α-lactose monohydrate and acetylsalicylic acid were comminuted in both devices. The results were evaluated by sieve analysis. In the jet mill, an impact angle of 60°

was used. Fig. 4.15 shows the determined breakage probabilities as a function of the specific kinetic energy for both devices and both materials. As can be seen, the measured breakage probabilities follow a Weibull distribution according to Eq. (2.2) in all cases. Data for the same substance coincide quite well for the two devices, the different behaviour of acetylsalicylic acid and lactose is obvious. The material parameters obtained from these data are shown in Table 4-2. Also given in this table are the standard deviations σ that describe the deviations of the experimental data from the model. From these numbers it becomes clear that the data points obtained from the jet mill scatter stronger than the data from the Schönert mill, but the results from both devices are comparable. The different impact angles (90° vs. 60°) do obviously not influence the comminution result in the case of the two tested substances.

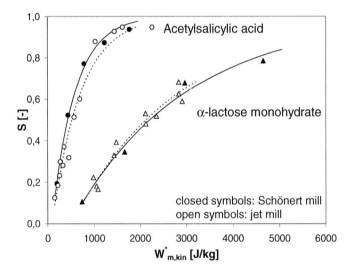

Fig. 4.15 Comparison of the comminution results for the two impact devices (evaluated by sieve analysis method)

Table 4-2 Material parameters obtained from evaluation by sieve analysis

Substance	Apparatus	f^{*}_{Mat} [kg/Jm]	$x \cdot W^{*}_{m,min}$ [Jm/kg]	σ [-]
α-lactose monohydrate	Schönert mill	2.23	0.0860	7.3%
	Jet mill	2.38	0.0916	13.2%
Acetylsalicylic acid	Schönert mill	4.67	0.0351	3.3%
	Jet mill	3.78	0.0392	15.7%

4.2.4. Breakage behaviour of different substances

In the following, comminution results for eight organic substances are presented. For a test of the quality of the population balance model, the breakage probabilities obtained from laser diffraction (LD) data are compared directly to those obtained from sieve analysis. In Fig. 4.16, data for seven of the examined substances are displayed (for Compounds A and B, no sieve analysis was done). The fit qualities for the LD data are given in Appendix D. The full line indicates that the result from laser diffraction is equal to the sieving result, the dotted lines indicate deviations +/- 10 %. It can be seen that most data points lie close to the equality line. In most cases, the data from laser diffraction are indeed a few percent higher than the data from sieving, as was expected in Chapter 4.1.4. Nevertheless, this deviation is negligible. Although the population balance using the Broadbent-Callcott distribution (Eq. (3.6)) yields a relatively poor fit quality in the region of fine particles (compare to Fig. 3.5), it obviously yields reliable results for the breakage probability. This can be explained because, for the determination of breakage probability, the decisive part of the distribution curve is the region of the coarser particles, in the size range of the feed material, and not the fines. The population balance model works sufficiently well to describe the size distribution in that part, i. e. it can describe the distribution of particles that disappear due to comminution, and this means, the breakage probability.

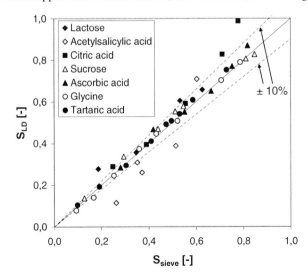

Fig. 4.16 Comparison of breakage probabilities determined from sieving and from laser diffraction (LD)

Exceptions to this are acetylsalicylic acid and some data points for lactose and citric acid. Different aspects may be considered to explain this outcome: the result of a particle size measurement - by sieving or by laser diffraction - is always dependent to some extent also on the particle shape. The plate-like shape of the acetylsalicylic acid crystals may have affected the result, as compared to the other substances that were closer to a cubic or spherical shape. Another possible influence is a different breakage mechanism for different substances. This may also have influences on the shape and the size distribution of the fragments and thus also on the size measurement and on the subsequent determination of the breakage probability. Additionally, also a relatively high amount of breakage in the nozzle may have skewed the results for easily-breaking substances, especially with the old nozzle design which was used for ASA, lactose and citric acid.

To check which effects these uncertainties have on the determination of the material parameters, the results for acetylsalicylic acid and lactose are compared as shown in Fig. 4.17. It can be seen from this figure that within a certain scatter, the results are still comparable, but the fit lines are somewhat shifted. This shift will mainly have consequences on the value of $x \cdot W^*_{m,min}$, since $W^*_{m,min}$ is represented by the intercept with the x-axis at $S = 0$. Generally, $x \cdot W^*_{m,min}$ is more susceptible to errors due to data scattering than f^*_{Mat}.

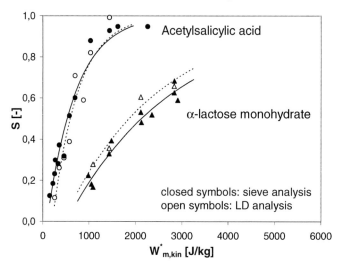

Fig. 4.17 Comparison between population balance evaluation (LD analysis) and sieving analysis

The breakage parameters obtained from both evaluation methods are given in Table 4-3. It can be seen that a noticeable deviation of f^*_{Mat} values exists for all substances. This is not only a result of data scattering, but also from another effect: the two analysis methods measure different equivalent particle diameters, which directly affect the results when applying Eq. (2.2): for the sieve analysis, the median value of the initial sieve fraction was used for the calculation. For evaluating LD data, the x_{50} value of the sample before milling, as measured by LD, was used. These values differ by factors between 1.13 (acetylsalicylic acid; compare to Table 4-3) and 1.32 (lactose). Using an x value increased by a factor 1.32 for the curve fitting of the breakage probabilities results in a $x \cdot W^*_{m,min}$ value increased by the same factor, while f^*_{Mat} decreases by that factor. In fact, the ratios of f^*_{Mat} from sieving versus f^*_{Mat} from LD lie within the same range (see Table 4-3). This explains the differing values for f^*_{Mat}. Exceptions to this are acetylsalicylic acid, where possible reasons have been discussed before, and citric acid, where at high impact velocities also deviating breakage probabilities were obtained, probably for the same reasons. Also, the results for $x \cdot W^*_{m,min}$ scatter stronger (especially for acetylsalicylic acid and glycine), which is a consequence of the generally higher sensitivity of this parameter.

Table 4-3 Material parameters determined from single particle comminution experiments ("LD"=laser diffraction, i.e. population balance analysis)

Substance	Results from LD			Results from sieving			ratio
	f^*_{Mat} [kg/ Jm]	$x \cdot W^*_{m,min}$ [Jm/kg]	σ [-]	f^*_{Mat} [kg/ Jm]	$x \cdot W^*_{m,min}$ [Jm/kg]	σ [-]	$x_{50,LD}$ /x_{sieve} [-]
α-lactose m.h.	2.04	0.103	5.9%	2.38	0.092	13.2%	1.32
Acetylsal. acid	3.76	0.101	17.8%	3.78	0.039	15.7%	1.13
Citr. acid m.h.	11.52	0.073	17.0%	8.76	0.054	8.1%	1.13
Compound A	0.60	0.657	1.0%	-	-	-	-
Sucrose	3.07	0.059	12.9%	3.79	0.056	6.6%	1.28
Asc. acid	6.26	0.006	8.3%	7.51	0	5.3%	1.26
Glycine	1.95	0.027	3.9%	2.22	0.014	2.6%	1.17
Tart. acid	1.21	0.098	4.5%	1.38	0.073	4.2%	1.21

m.h. = monohydrate

For Compound B, no values could be specified for the breakage parameters. This substance consists of rod-shaped particles that break very easily at all tested impact velocities. Therefore, a breakage probability of 100% is obtained in all experiments, and the Weibull model cannot be applied properly. This shows again that the particle shape also has an influence on the results.

In Fig. 4.18, the breakage probabilities for all substances are plotted over the product $f^*_{Mat} \cdot x \cdot (W^*_{m,kin} - W^*_{m,min})$. The data come to lie well within the 15% confidence limits of the mastercurve that was deducted by Vogel and Peukert (2003). The fact that the breakage behaviour of all different substances can be described by this mastercurve confirms that this concept is also valid for the material class of molecular organic crystals, which covers the majority of pharmaceutical powders.

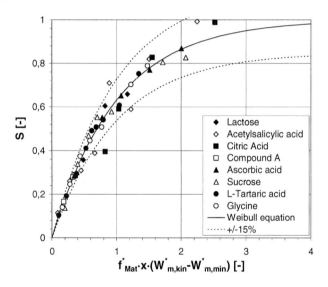

Fig. 4.18 Mastercurve for the breakage probability

4.2.5. Influence of feed size and impact angle

To check for the influence of particle size on comminution, different sieve fractions of α-lactose monohydrate were tested, the smallest fraction being between 45 and 63 μm. The breakage probabilities obtained from LD data are displayed over the product $x \cdot W^*_{m,kin}$ in Fig. 4.19. This way of plotting eliminates the size dependency of the parameter $W^*_{m,min}$, i.e. if the parameters $x \cdot W^*_{m,min}$

and f^*_{Mat} are constants, then all data points should lie on the same curve. In fact, the data do not scatter to a higher extent than found before for each single feed size fraction, and they show no tendencies regarding different size fractions. Determining the material parameters from fitting these data for all size ranges yields $f^*_{Mat} = 1.79$ kg/Jm and $xW^*_{m,min} = 0.0908$ Jm/kg. These values deviate from the values determined from only one size fraction by about 14 %, which is within the limits of error as will be discussed in the following section. Therefore the conclusion can be drawn that f^*_{Mat} of lactose is not dependent on particle size in the range down to 45 µm. Vogel and Peukert (2003) had already examined glass spheres down to 100 µm and also found no size dependency of f_{Mat}. This seems to apply also for crystalline organic substances.

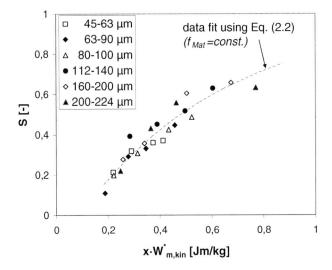

Fig. 4.19 Breakage probabilities for different sizes of α-lactose monohydrate (LD analysis)

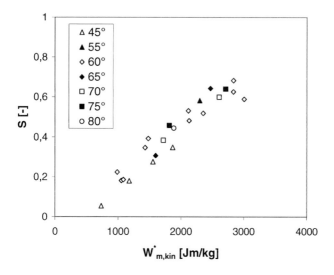

Fig. 4.20 Breakage probabilities of α-lactose monohydrate at different impact angles

Also, the question is addressed how the impact angle affects comminution. α-lactose monohydrate with a particle size of 160-200 μm was comminuted at different impact angles between 45 and 80°. The results are shown in Fig. 4.20. For the kinetic energy, only the component normal to the impact plate was considered. From the graph, no trend can be seen with changing impact angle, all data points lie within the general scatter of the data. The dependency of the breakage probability on the impact angle was studied also by several other authors. Menzel (1987) varied the impact angle of limestone between 30° and 90° and found practically constant results for impact angles of 75° and above; at lower angles, a coarser product was obtained. Salman et al. (1995) studied oblique impacts of 5.15 mm aluminium oxide spheres. They report a decrease of breakage probability at angles below 50°, while it varies only slightly at angles between 50° and 90°. In both cases, these authors considered the entire particle velocity, not only the component normal to the impact plane. Still the results presented here are comparable with their results: at 75°, the energy component normal to the impact plane is reduced by only 7 % compared to a 90° impact, i. e. we are talking about small deviations here anyway. Also, a different breakage mechanism may react differently to a changed impact angle; this cannot be taken into account, since different materials were studied by these authors.

4.2.6. Error discussion for comminution in the air jet device

The biggest inaccuracy in the comminution results from the air jet mill arises from the determination of impact speed. The particles do not impact all at the same velocity. Even if narrow sieve fractions of feed material are used, the particle diameter is not exactly constant, and therefore the acceleration of the particles is different. The measured particle velocity distributions were found to have a standard deviation up to 10 %. Since the velocity goes into the impact energy with the power of 2, standard deviations of up to 20 % are obtained. Another possible error arises from the fact, that the particle velocity is measured about 2 mm above the impact plate, and not directly on the plate. As already pointed out in Chapter 4.2.2, the velocity in this point deviates from the real impact velocity no more than 5 % for particles larger than 25 μm. Also, the deviation of the impact angle of a particle of that size would result in less than 5 % deviation of the impact velocity. These deviations are lower than the scattering of the velocity measurements, therefore these errors should be negligible against the error arising from the measured velocity distribution.

For the determination of the breakage probability, the accuracy of the sieve analysis method can be given as ± 4 %. Comparing the breakage probabilities from LD analysis to those obtained from sieve analysis, relative deviations below 10 % are found for most substances, which correspond to absolute deviations of less than 5 % in most cases, only for acetylsalicylic acid, deviations up to 20 % were found. The standard deviations of the data fits using Eq. (2.2) are given in Table 4-2 and Table 4-3. Deviations of up to 19.6 % are obtained. These deviations are a consequence of the scattering of the experimental data explained above, and they lie within the same range as the experimental standard deviations.

One more remark about the use of the Broadbent-Callcott equation for the modelling of the LD data: a better representation of the distribution of the fine particles surely could be achieved by using a more detailed breakage function. As of now, the breakage function is modelled with a relatively simple equation. This bears the advantage that no additional parameters except S need to be adjusted. One has to keep in mind that, when using more detailed breakage functions more parameters have to be introduced, that theoretically can differ not only from substance to substance but may also change already for different impact velocities. Therefore, care has to be taken when chosing another model for the breakage function. The present, simple model has shown its usefulness in determining the breakage probability for most of the examined substances.

4.2.7. A general model for breakage functions

So far, the breakage probability of different organic materials has been discussed. Now, also the breakage function shall be considered. For this, the size distributions after breakage, as obtained by laser diffraction, were studied in detail. The results shown in this section are also published by Meier et al. (2009a). In addition to the organic materials studied so far, three more materials were studied that were already comminuted by Vogel and Peukert (2003): limestone, ammonium sulphate and an epoxy coating powder. Since the impact velocities were chosen in the range of incomplete comminution (i. e. the breakage probabilities were below 90 %), the collected samples still contain non-broken particles. In order to characterise the fragment size distribution, i. e. the distribution of the comminuted particles, this portion has to be separated. The evaluation procedure will be explained in the following with the example of glycine.

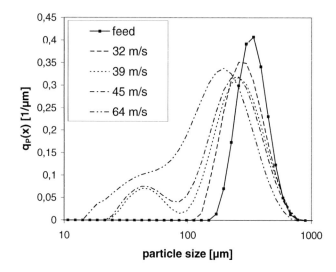

Fig. 4.21 Particle size distributions of glycine after impact comminution

Fig. 4.21 shows the size distributions of glycine, impacted at different velocities. The features presented here can be observed with all substances studied: the density distribution reveals two modes. With increasing impact speed, the peaks are getting broader, and the coarser one moves towards finer particle sizes, until the peaks unite into one broad distribution.

These distributions have to be interpreted as a mixture between broken particles (their size distribution being the breakage function), and non-broken particles that still have the same distribution as the feed material. For every size class Δx, the distribution of this mixture can be described by

$$q_P(x)\Delta x = S \cdot q_B(x)\Delta x + (1-S) \cdot q_F(x)\Delta x \qquad (4.3)$$

Here, q_P represents the density distribution of the product as measured after milling, q_B represents the breakage function, q_F the distribution of the feed material and S the breakage probability. With this equation, the fraction of non-broken particles can be separated from the broken particles, since the breakage probability S is known. Thus, a breakage function can be calculated for each impact speed. This procedure is visualised in Fig. 4.22. This figure shows how the measured product distribution consists of a fraction of feed material that has not been broken on impact, and of a fraction of broken particles, that has the size distribution of the breakage function. As can be seen from Fig. 4.22, the thus determined breakage function has a third mode in the region of coarse particles; in some cases, even negative values are obtained. This contradicts the logical assumption that the fragments obtained after breakage are smaller than the original, non-broken feed particles. These effects obviously are introduced by

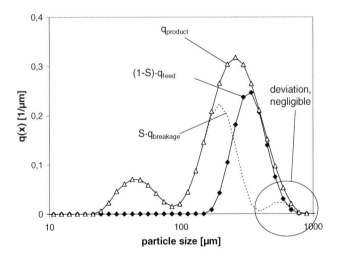

Fig. 4.22 Illustration of the process to determine the breakage function (example: glycine at 39 m/s); the evaluation combines two different methods of size determination, which may lead to deviations from ideal bimodal breakage functions

the calculation procedure as a consequence of small variations in the measured data. The evaluation combines two different methods for size determination, which can also cause some deviations. This third mode represents between 5 and 10 % of the entire distribution, in a few cases up to 15 %. Still it seems reasonable to neglect this third mode and assume a bimodal form of the breakage function.

At higher impact speeds, sometimes another mode at very fine particle sizes is obtained. In this case, this mode represents up to 3 % of the entire distribution for most substances, for lactose, up to 7 % are reached. Because these particles contribute only a small amount to the entire sample mass, and in order to keep the evaluation simpler, these modes are ignored as well. These finest size fractions may become relevant at higher impact velocities, i. e. under process conditions where the feed particles are destroyed completely.

The thus determined breakage function curves were fitted with bimodal logarithmic normal distributions. This type of distribution was chosen because, considering the points discussed in Section 2.1.2, logarithmic normal distributions are suited best to model the breakage functions. We did not truncate these functions, as was done by Schubert and co-workers, in order to keep the number of parameters as small as possible and thus to keep the evaluation simple. As was noted already by Crabtree et al. (1964), bimodal distributions appear in the case of what these authors call "attrition grinding", which is considered to take place when there is insufficient energy or too inefficient loading to produce complete fracture of the particle. This is the case in our experiments, since the stressing conditions were such that breakage probabilities smaller than 90 % are obtained.

According to the considerations above, the following equation for a bimodal logarithmic normal distribution was used to model the breakage functions:

$$q_B(x) = \frac{a_1}{x\sigma_A\sqrt{2\pi}}exp\left\{-\frac{1}{2}\left[\frac{ln\left(\frac{x}{x_A}\right)}{\sigma_A}\right]^2\right\} + \frac{1-a_1}{x\sigma_B\sqrt{2\pi}}exp\left\{-\frac{1}{2}\left[\frac{ln\left(\frac{x}{x_B}\right)}{\sigma_B}\right]^2\right\} \quad (4.4)$$

Here, x_A and x_B are the median values of the peaks, σ_A and σ_B are the standard deviations. The index "A" will be used for the peak of the finer particles, "B" for the peak of the coarser particles. The factor a_1 denotes the "mixing ratio" of the two peaks. Based on this equation, a data fit was made to the calculated fragment distribution with the aid of the software SigmaPlot 9.0. An example for such a fit is shown in Fig. 4.23. By performing this data fit, we obtain a set of five parameters for each single experiment. The aim of the following evaluation is to identify the trends how these parameters change with different influencing factors, i.e. impact energy, particle size and material. Based on this, a universal breakage function will be determined.

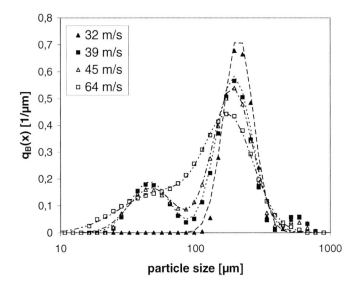

Fig. 4.23 Bimodal log-normal distributions fitted to the calculated breakage functions of glycine; points = calculated breakage function; lines = fit of bimodal LND

For a first test of the breakage functions as they are derived by this method, the median diameters of the functions are plotted against the dimensionless stressing parameter. Fig. 4.24 shows a comparison of the data obtained in this work to the data by Vogel and Peukert (2003). In the Vogel and Peukert data, two regimes can be distinguished: up to a dimensionless stressing parameter $f^*_{Mat} \cdot k \cdot x \cdot W^*_{m,eff}$ of about 2, the median diameter $x_{50,3}$ decreases only slightly. In this region, breakage probabilities below 90% are achieved. In the second regime, breakage probabilities lie near 100%, and the decrease of $x_{50,3}$ is much stronger. Because the experiments in this work were mainly aimed at determining the breakage parameters f^*_{Mat} and $x \cdot W^*_{m,min}$, breakage probabilities smaller than 90 % are achieved in most experiments, and the data points obtained in this work lie mostly in the first region of the graph. Moreover, these data points lie slightly below Vogel's and Peukert's points. Most probably, this is caused by the different evaluation methods, since Vogel and Peukert analyzed sieving data and not laser diffraction data. Note that three of the materials studied here were also studied by Vogel and Peukert. The results for these substances (obtained from evaluation of laser data) coincide very well with the other materials of this work, while in the case of the sieve evaluation done by Vogel and Peukert, the data lie within those of their other substances. Accordingly, the slight deviation of the results must be due to the evaluation method rather than due to a material-specific influence.

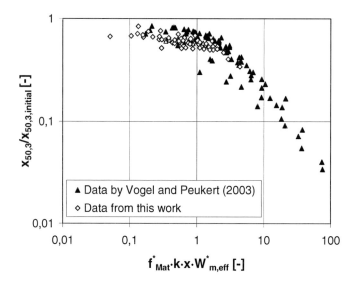

Fig. 4.24 Comparison of Vogel and Peukert's (2003) data and data from this work: median diameters of measured fragment size distributions

In the following, the breakage functions and their parameters (x_A, x_B, σ_A, σ_B, a_1) shall be characterised in detail. For this, we will first have a look at the influence of the feed particle size on the parameters; later, different materials will be compared. Data from six different particle size ranges were available for α-lactose monohydrate. The median diameters are made dimensionless by dividing them by the median diameter x_0 of the (monomodal) size distribution of the feed. The same is done with the standard deviations σ, they are divided by σ_0 of the initial size distribution.

Fig. 4.25 shows the results for the standard deviation of the coarser mode, σ_B, for different particle sizes. On the left hand side, the dependency on the effective impact energy is shown, and it can be observed, that σ_B is slightly higher for coarser feed material. A replot of the data over the product of particle size and effective impact energy, as shown on the right hand side of Fig. 4.25, yields a coincidence of all data points on the same curve. This shows that the parameter σ_B depends on the initial particle size of the feed.

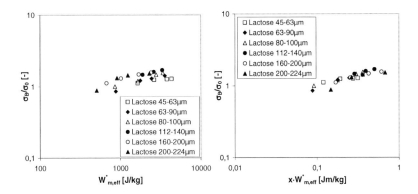

Fig. 4.25 Standard deviation of the coarse peak for different size fractions of lactose

In the same way, also the other four parameters are plotted over $x \cdot W^*_{m,eff}$ in Fig. 4.26. With the exception of the parameter a_1, where a strong scattering of the data can be seen, the results for the other parameters come to lie on one curve. In the graph for σ_A, the data set for the smallest sample size, 45-63 µm, lies noticeably higher than the other data points. This is a consequence of the evaluation procedure: the density distributions of all lactose samples exhibit a third mode in the region of very fine particles. As described above, this mode is neglected in order to keep the evaluation more simple. With the smallest feed size, however, this mode overlaps with the finer mode "A" of the bimodal distribution, and the overlapping part is cut out for the data fit. As a consequence, less data points are available for fitting Eq. (4.4) to the finer mode, and the data fit becomes more difficult. In fact, the fit quality in the region of the finer mode is not as good as for the other lactose samples. If σ_A is manually kept constant at a smaller value, an almost equally good data fit is obtained, with the other parameters only slightly changing. Therefore, the obtained values are not really reliable in this case, and the data set for the 45-63 µm lactose samples will be disregarded in the following comparison with other substances. So far, the results show that a size dependency exists for at least three of the parameters

(σ_A, σ_B and x_A) of the bimodal breakage function. For x_B, the values were found to be almost constant, therefore it is hard to make a definite statement from those data. Also for the parameter a_1, the size dependency cannot clearly be given because still a strong scatter of the data persists.

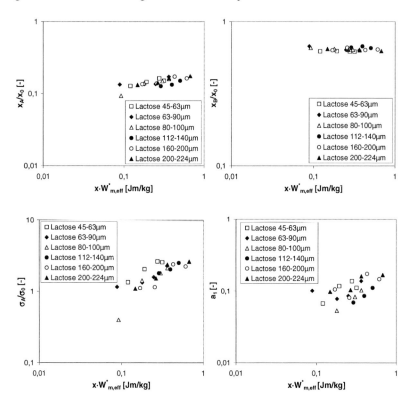

Fig. 4.26 Parameters of the breakage function; plotting the data vs. $x \cdot W^*_{m,eff}$ shows the size dependency of the parameters

Next, we will compare the breakage function parameters of different materials. Fig. 4.27 shows the normalised standard deviation σ_B of the coarser peak. In the upper graph, the data are plotted over the product $x \cdot W^*_{m,eff}$, i.e. the size dependency is already considered. Still, it can be clearly seen that the standard deviation is different for different materials. Therefore, the material properties need to be accounted for as well. This is done in the lower graph of Fig. 4.27, where the data are plotted over the complete dimensionless stressing parameter

$f^*_{Mat} \cdot x \cdot k \cdot W^*_{m,eff}$ (the number of impacts, k, however, is equal to 1 in all experiments done in this work). Now, all data points come to lie on the same curve. This again illustrates how the breakage relevant material properties, expressed by f^*_{Mat} and $x \cdot W^*_{m,min}$, influence the breakage function.

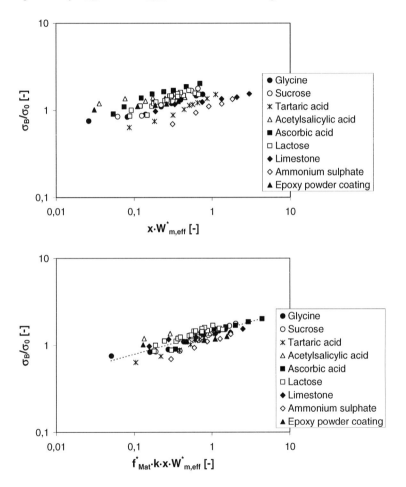

Fig. 4.27 Normalised standard deviations of the coarse mode - the standard deviation can be described by one single curve if the material properties (i.e. f^*_{Mat}) are properly accounted for

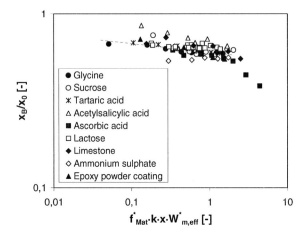

Fig. 4.28 Normalised median diameter x_B as a function of the dimensionless stressing parameter

The same evaluations are done for the other parameters of the approximated breakage function. Fig. 4.28 shows the variation of the normalised median diameter of the coarser peak. Here, a curve is obtained that is similar to the one in Fig. 4.24, where the median diameter of the entire breakage function is plotted. The diameter decreases only slightly up to a stressing parameter of about 2. Above 2, the decrease becomes markedly stronger (represented in this graph only by two data points). This indicates that the breakage functions might change if more energy is supplied.

Fig. 4.29 shows the normalised standard deviation σ_A of the finer peak. A similar material dependency as for σ_B is detected. In contrast to that, there seems to be no material dependency for the median diameter of the finer peak. Here, a better correlation is obtained if the material parameter f_{Mat} is not included, i.e. x_A/x_0 is plotted over the product of size and effective impact energy (see Fig. 4.30). Also in the case of the mixing factor a_1, the best correlation was found with the product $x \cdot W^*_{m,eff}$ (see Fig. 4.31). Still, the data scatter is relatively high, as was already observed with the lactose samples of different sizes. Therefore it can not be entirely excluded that the material has an influence on a_1, but if so, the influence is too small to be detected under the scatter of the data. Nevertheless, a size dependency of a_1 seems to exist.

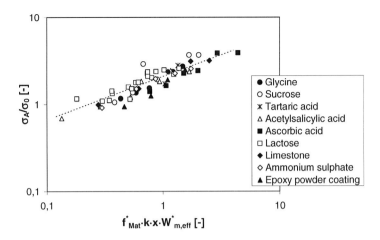

Fig. 4.29 Normalised standard deviations of the fine mode

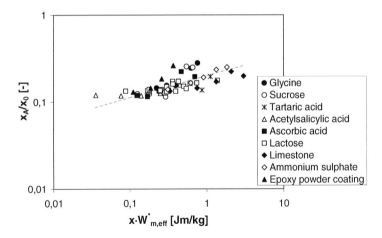

Fig. 4.30 Normalised median diameters of the fine mode; plotting over a material-independent parameter yields a better correlation

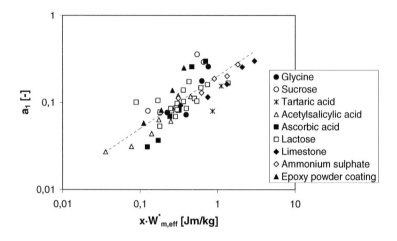

Fig. 4.31 Mixing ratio a_1 of the two modes of the bimodal breakage function

Based on the correlations presented above, empirical mathematical expressions can be given for the breakage function. The following equations correspond to the dashed fit lines indicated in the previous figures. The equations show that the complete breakage function can be predicted if the impact energy, the initial particle size and the breakage parameters f^*_{Mat} and $xW^*_{m,min}$ are known.

$$\frac{x_A}{x_0} = 0.2 \cdot \left(x \cdot W^*_{m,eff}\right)^{0.25} \tag{4.5}$$

$$\frac{x_B}{x_0} = 0.6 \cdot \left(f^*_{Mat} \cdot x \cdot W^*_{m,eff}\right)^{-0.05} \tag{4.6}$$

$$\frac{\sigma_A}{\sigma_0} = 2.1 \cdot \left(f^*_{Mat} \cdot x \cdot W^*_{m,eff}\right)^{0.5} \tag{4.7}$$

$$\frac{\sigma_B}{\sigma_0} = 1.4 \cdot \left(f^*_{Mat} \cdot x \cdot W^*_{m,eff}\right)^{0.25} \tag{4.8}$$

$$a_1 = 0.2 \cdot \left(x \cdot W^*_{m,eff}\right)^{0.6} \tag{4.9}$$

The data presented so far are based on experiments with impact angles of 60°. The question remains, whether the results are also valid for other impact angles. Therefore, additional comminution experiments were conducted in the single particle impact device after Schönert and Marktscheffel (1986), where an impact angle of 90° is realised (see also Chapter 3.1.1). Evaluation of the data was done in the same way as described above with the 60° impacts. One easily-breaking substance (acetylsalicylic acid) and one substance with intermediate breakage behaviour (glycine) were chosen for these experiments. With each substance, three experiments were done, chosing impact velocities leading to low, intermediate and high extents of breakage. The comparison of the parameters obtained at 60° and 90° impact is shown in Fig. 4.32. It can be easily seen that for all the parameters, the results coincide very well, within the same scattering as observed before. From this it can be concluded that the breakage function does not depend on the impact angle, only on the velocity component normal to the impact plane. At 60°, the tangential velocity component has no effect on the breakage function; however, it is still possible that the tangential component becomes more important at smaller impact angles, which has not yet been tested.

The presented results are valid for different kinds of material. However, Lecoq et al. (2003) classify various materials into different groups according to their breakage behaviour: brittle (glass, sand, polyamide, NaCl), ductile (PMMA), and "complex" ($Al(OH)_3$). This classification is based on the observation of various fineness criteria: size reduction ratio, weight percentage of particles finer than 40 μm, and specific surface area. Since these criteria are related to the fragment size distributions after breakage, this indicates that it is still possible that there are materials having breakage functions different from the ones shown here. However, this classification seems not to be applicable to general material classes such as "crystalline" or "polymeric materials", but rather is attributed to specific features of the material structure, as Lecoq et al. explain for the "complex" behaviour of $Al(OH)_3$ with its structure consisting of piled platelets.

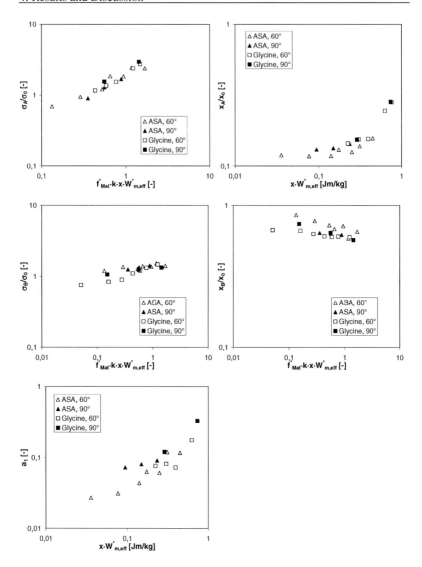

Fig. 4.32 Comparison of the breakage function parameters for impacts at 60° and 90°

4.3. Nanoindentation

4.3.1. Determination of hardness and Young's modulus

Fig. 4.33 shows the load on the sample, Young's modulus and hardness over the indentation depth for eight experiments on different lactose crystals (several 100 μm large). At very low loads, various influences affect the measured properties. The seemingly very high hardness and modulus in this region are explained by the so-called indentation size effect (ISE): during indentation, dislocations are created which are required to account for the permanent shape change at the surface. These dislocations are called "geometrically necessary dislocations" (GND), and they increase the effective yield strength and thus the local hardness of the material. The density of the GND decreases with increasing indentation depth, resulting in a lower hardness and Young's modulus (Nix and Gao, 1998). For soft and crystalline materials, a significant ISE can be expected (Fischer-Cripps, 2004), and is observed in most experiments. Several other factors also can influence the results especially at very low indentation depths, such as surface roughness, slipping or friction between tip and sample.

All these influences become less pronounced as the indentation depth is increased, and after a few 100 nm indentation depth, a level is reached where hardness and Young's modulus remain constant. Taking a close look at the load curves (see Fig. 4.33a), several small discontinuities in the curvature can be observed at higher loads. These are so-called "pop-in" events that happen when cracking occurs during loading. The biggest one of these pop-ins is marked with an arrow in Fig. 4.33a. Further, it can be seen, that there is still a slight decrease of hardness the deeper the indent gets (Fig. 4.33c). It seems that the failing of the material at higher loads leads to a decreasing hardness. This can be even better illustrated with the example of acetylsalicylic acid. Fig. 4.34a shows the load over the indent depth for an indentation in acetylsalicylic acid. The curve displays several pronounced pop-in-events; while in the load curves of lactose, just small discontinuities are seen, now large steps are visible. Fig. 4.34b shows a close view of both the load and the hardness during one indentation. Simultaneously to the pop-in, a stepwise decrease of hardness can be observed. Obviously, the cracks that are now present in the material lead to a decrease of the measured hardness.

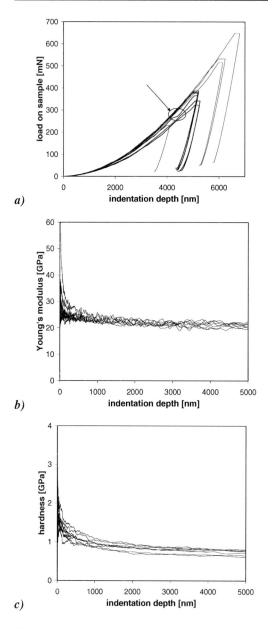

Fig. 4.33 (a) Load, (b) Young's modulus and (c) hardness during indentation of α-lactose monohydrate

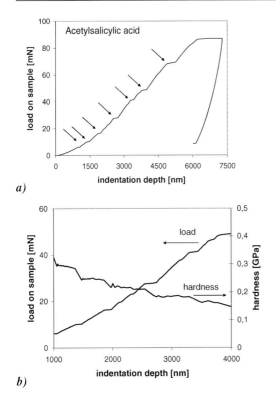

a)

b)

Fig. 4.34 (a) Pop-in-events during indentation of acetylsalicylic acid, connected to material failure; (b) decrease of measured hardness simultaneous to the pop-in steps

This example illustrates the difficulties arising when determining the hardness of a material. The question is, which is the correct hardness value that can be used for a consistent comparison of different materials? Artefacts coming both from material failure and from the indentation size effect should be avoided. As a compromise, hardness and Young's modulus are determined in this work by averaging the measured values of E and H in the depth region starting where a constant level is reached until marked cracking starts. For most substances, this was at indentation depths around 2000 nm (see Table 4-4). Although these depths are sufficient to obtain hardness and Young's modulus of the materials, higher indentation depths were chosen for the experiments, because in most cases higher loads are required to induce cracks large enough for the determination of fracture toughness. The measured values for E and H of the tested substances are given in Table 4-4. The standard deviations lie typically around 5 % for Young's modulus and around 10 % for the hardness. The values range from 0.2 to 1.4 GPa for the hardness and from 5 to 45 GPa for Young's modulus. This represents a medium range between softer materials such as polymers and more rigid materials such as metals or ceramics.

Table 4-4 Material properties H and E and their standard deviations σ, obtained from nanoindentation

Substance	H [GPa]	σ (H) [-]	E [GPa]	σ (E) [-]	E/H [-]	Depth range for data aquisition [nm]
α-lactose m.h.	0.869	7.3%	21.44	4.5%	24.7	1500..2000
Acetylsalicylic acid	0.222	11.8%	5.44	5.1%	24.5	1300..2000
Citric acid m.h.	0.608	12.4%	16.38	8.3%	40.0	3000..4000
Compound A	0.248	5.4%	9.93	9.8%	34.4	4000..5000
Compound B	0.340	25.0%	10.70	9.6%	31.5	500..1000
Ascorbic acid	1.003	5.6%	34.46	5.1%	23.7	1000..2000
Sucrose	1.013	12.6%	23.97	6.3%	32.0	2000..3000
Tartaric acid	1.354	3.5%	43.34	1.5%	27.7	2000..2500
Glycine	0.852	4.1%	23.56	4.6%	26.9	1300..2200

m.h. = monohydrate

Some of the substances studied were also assessed with indentation techniques by other authors. Sucrose was studied by Duncan-Hewitt and Weatherly (1989) who employed Vickers indentation at a load of 147 mN, determining the hardness from the dimensions of the indent. They found a hardness of 0.645 ± 0.035 GPa and a Young modulus of 32.3 GPa. Zügner (2002) used the Oliver and Pharr method for the evaluation of indentations of several 100 nm penetration depth. He determined $H = 1.2 \pm 0.15$ GPa and $E = 29 \pm 3$ GPa for sucrose. These values are about 20% higher than our values, which most probably is due to the different indentation depth used, as it was discussed in the previous paragraph. The lower value by Duncan-Hewitt and Weatherly can be explained by their higher indentation load which was about twice as high as it was in our measurements at a depth of around 2 μm. α-lactose monohydrate was assessed by nanoindention by Fagan et al. (1996) who determined $H = 0.67 \pm 0.23$ GPa and $E = 26 \pm 7.1$ GPa at a maximum load of 50 mN, which agrees well with the data presented here. Zügner et al. (2006) found $H = 1.1$ GPa and $E = 23.7$ GPa for lactose, which is, considering the smaller penetration depth, also in good agreement with the data obtained in this work.

4.3.2. Fracture toughness

Choice of model

As already pointed out in Chapter 3, many models have been developed to obtain fracture toughness from indentation experiments, mostly derived for ceramic materials, which typically have a hardness between 10 and 20 GPa and a fracture toughness between 1 and 10 MPa$\sqrt{m}$. Therefore it is not clear, whether these models are also applicable for softer materials, namely pharmaceutical substances. As already mentioned, most of the models assume that $P/c^{3/2} = $ const. (Eq. (2.24)). This relationship comes from fracture mechanical modelling of well-developed halfpenny cracks. It was shown by several authors (Duncan-Hewitt and Weatherly, 1989; Prasad et al., 2001; Taylor et al., 2004a), that Eq. (2.24) is valid for various pharmaceutical substances. Also in this work, it was found that plots of ln c versus ln P yield slopes between 0.57 and 0.78 for most substances, which can be considered close enough to the exponent 2/3 required for Eq. (2.24) to be valid (see Fig. 4.35). Exceptions to this are acetylsalicylic acid with a slope of 1.9 and ascorbic acid with a slope of 1.11. This is probably attributed to the anisotropies in the crystals, which were observed in the indentation experiments for these two substances. Also, lactose displays a slope of 0.98. This value is already a bit far from the required 2/3, but might be caused by a relatively big scatter of the experimental data that is also reflected in the fracture toughness values for lactose. With citric acid, only one load was applied with a microhardness tester, therefore the P-c-dependency could not be tested. Most of the studied substances fulfill the basic requirement for the fracture toughness models. Also acetylsalicylic acid and ascorbic acid will be included in subsequent evaluations; as will be shown, most of the models yield similar results, irrespective of the true crack system.

For ceramic materials, fracture toughness data from indentation can be directly compared to results from conventional fracture toughness determination, i.e. strain experiments using notched specimens of defined geometry, so-called SENB experiments (SENB = single edge notched beam; the obtained value is the stress intensity factor K_{IC}, the index I standing for mode I fracture, i.e. when plain strain is applied). For organic substances, single crystals that are large enough for these tests are hard to obtain. With pharmaceutical powders, such tests often are done using compacted powder specimens of different porosity, and extrapolating the results to zero porosity (Roberts et al., 1993). But, as pointed out by Larsson and Kristensen (2000), it is questionable whether results of these experiments with powder compacts can be transferred to single crystals of organic substances. In fact, data were shown to vary strongly, depending on

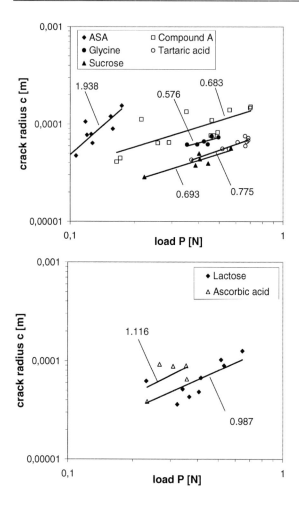

Fig. 4.35 Correlation between crack radius and applied load; slopes are indicated in the graph; for reasons of legibility, the data are shown in two separate graphs

particle size and specimen dimensions (see e. g. Bin Baie et al., 1996; Podczeck, 2001). Fagan et al. (1996) showed that results from indentation and from beam bending can be very different, depending on the beam quality. Literature data for α-lactose monohydrate and acetylsalicylic acid obtained from SENB experiments (Roberts et al., 1993; Podczeck, 2001) are far higher than the values given here that were obtained from indentation. However, measurements performed on single crystals should provide a more realistic picture, as far as the material characteristics are concerned.

Therefore, the results shown here could not be compared to literature values from SENB experiments. Instead, the different model equations were compared relative to each other. The quantities that enter these models usually are hardness, Young's modulus, applied load, the indent diagonal and the crack length. With these quantities measured, all 19 model equations collected by Ponton and Rawlings (1989a, 1989b) were used to evaluate the fracture toughness of the tested materials. Fig. 4.36 shows the averaged fracture toughness values for sucrose, obtained with the different model equations. Note that the first 15 models are based on the assumption of half-penny crack systems, while the models 16-19 are based on Palmqvist crack systems. The results differ significantly: the majority of equations yields results between 0.09 and 0.14 MPam$^{1/2}$. The first two models yield markedly lower values, probably due to the fact that these models are based on linear-elastic fracture mechanics as opposed to elastic-plastic behaviour in the other models. On the other hand, no significant difference can be seen between results of halfpenny crack equations and Palmqvist crack equations. This means that models using different assumptions yield similar results, a fact that was already observed by Ponton and Rawlings (1989b). The same observations can be made with the other substances studied.

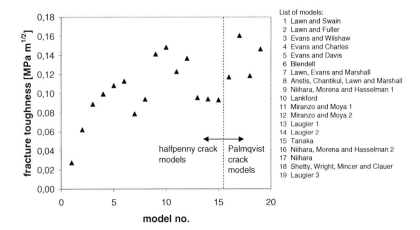

Fig. 4.36 Fracture toughness for sucrose, obtained with different equations; the numbering of the model equations is according to the order as they appear in Ponton and Rawlings (1989a)

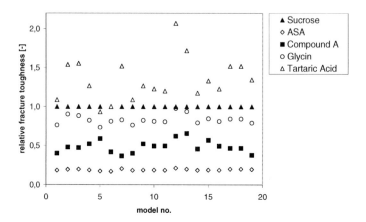

Fig. 4.37 Relative fracture toughness values $K_C/K_{C,sucrose}$ for different organic substances; the numbering of the model equations is according to the order as they appear in Ponton and Rawlings (1989a)

In the next step, values of the fracture toughness of different materials are compared relative to each other for each model. This is graphically represented in Fig. 4.37, where the fracture toughness values relative to the fracture toughness of sucrose $K_{C,rel} = K_C/K_{C,sucrose}$ are shown. Sucrose was chosen as the standard for this comparison. It can be seen that, irrespective of the absolute values and of the crack system, the relation between different materials remains similar for most of the equations. These observations are an indication that, for an empirical comparison of the fracture toughness of different organic materials, almost any of these equations can be used. Since models derived for different crack systems lead to similar relations, it seems that the models are also applicable for substances that do not fulfill the halfpenny crack criterion ($P/c^{3/2}$ = const.). In fact, the variation of relative values is not stronger for ASA and ascorbic acid than for other substances.

For the purpose of this work, which is to characterise correlations between the fracture toughness and the milling performance, a model is needed that provides also a good quantitative ranking of different substances. Therefore, a model has to be chosen that is capable of giving good results especially for softer materials (recall that the hardness and fracture toughness of molecular organic crystals are about one order of magnitude lower than those of ceramics). For ceramics, the relation $P/c^{3/2}$ = const. was found to be valid for large ratios of crack length c to indent diagonal a. Therefore, in many of the models boundary conditions exist requiring a minimum value for c/a. Often, c/a > 2 is required, sometimes even c/a > 2.5. For several of our substances, c/a values around 2 to 2.5 were obtained with the loads that could be applied with the MTS system (max. 700 mN), which is already close to this limit (see Table 4-5). Therefore, a model should be used without such a requirement in c/a. It was decided to use the Evans-Davis equation (Eq. (2.28)) for evaluation, because it can provide a good quantitative correlation also for softer materials and lower c/a values. This equation was already introduced in Section 2.2. This choice is also supported by the evaluation made by Ponton and Rawlings (1989b). These authors evaluated the equations regarding both their "correlation ability", i.e. the ability to correlate K_C and K_{IC}, and their "ranking ability", i.e. their ability to rank different ceramic materials according to their fracture toughness. Their final recommendation for the best combined ranking and correlation ability (i.e. a quantitative ranking of different materials) is the use of the equations by Evans and Charles (1976) (model no. 4 in Fig. 4.36), Evans and Davis (1979) (model no. 5) or Blendell (1979) (model no. 6). The Blendell model essentially is also a curve fit to the Evans and Charles data, and in fact, similar data are obtained from the Blendell model and the Evans and Davis model.

Own measurements of fracture toughness

Indentation depths between 5000 and 13000 nm were used in order to induce cracking in the crystals. It was observed that some materials required significantly higher loads than others before cracking occurred. Some examples of indents are shown in Fig. 4.38. For lactose, tartaric acid and citric acid, radial cracks are observed; for sucrose, both radial cracks and a lateral chipping is observed, indicating a superposition of radial and Palmqvist cracking. In the case of acetylsalicylic acid and ascorbic acid, a very strong anisotropy could be seen - cracks develop always in the direction of a certain crystal orientation, therefore not necessarily emerging from the corners of an indent, but from the center of the indent. With glycine and compound A, lateral cracks in addition to radial cracking were observed. However, evaluation of these cracks was done applying the same model as for the other cracks, since the kind of crack system should not have much effect on the evaluation, as was discussed in the previous section. In order to get a statistically reliable picture, at least five indentations on different crystals have been evaluated for each material. The results are given in Table 4-5.

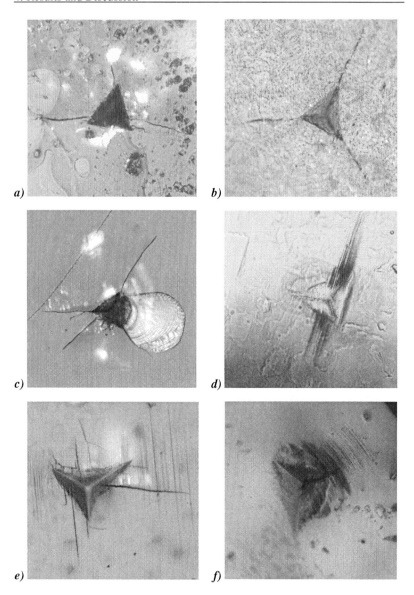

Fig. 4.38 Indents and resulting cracks in various materials: (a) α-lactose monohydrate, (b) tartaric acid, (c) sucrose, (d) acetylsalicylic acid, (e) glycine, (f) Compound A

Table 4-5 Fracture toughness of the examined materials (evaluation after Evans-Davis, 1979)

Substance	K_c [MPam$^{1/2}$]	$\sigma(K_c)$ [-]	c/a ranging from	# of crystals	# of indents
α-lactose m.h.	0.0908	32.1%	1.8-4.5	4	9
Acetylsalicylic acid	0.0211	35.8%	1.6-4.3	6	9
Citric acid m.h.[*]	0.0462	12.6%	1.7-2.6	8	8
Compound A	0.0518	34.3%	1.8-2.8	4	14
Compound B	0.0144	29.3%	1.9-3.0	3	8
Ascorbic acid	0.0776	46.2%	2.8-6.2	2	5
Sucrose	0.1083	30.8%	1.5-3.5	7	10
Tartaric acid	0.1666	8.3%	2.5-3.5	2	10
Glycine	0.0978	8.4%	1.9-2.8	2	8

m.h. = monohydrate

[*]*evaluated from Vickers indentation*

With citric acid monohydrate, evaluation of the crack length was somewhat more tricky: often, no cracks or even indents could be found after the indentation. It was then observed that indents as well as cracks are formed, but disappeared after some time - obviously some kind of "healing" or recrystallisation took place. This is illustrated in Fig. 4.39, where it can be seen that cracks get smaller and disappear within several minutes when kept at room conditions at intermediate humidities. Presumably, the water contained in the ambient air lead to a recrystallisation or caking of the crack surfaces. The indentation experiments were typically done at room temperature and at relative humidities between 40-70 %, the room climate was not controlled. Thus the cracks induced by indentation could heal out, helped by the humid environment, before they were measured. Basically, changes of crack length after indentation can happen with any material, e. g. Ponton and Rawlings (1989) mention an environmentally assisted slow crack growth for silicate glasses. In this work, citric acid monohydrate was the only substance that was found to have crack lengths changing after indentation. As a consequence, pictures of the indent had to be taken as soon as possible. An accurate determination of the crack lengths is not possible with the optical unit of the MTS Nanoindenter, therefore samples always have to be transferred to another microscope with better possibilities in terms of magnification and illumination. But the dismantling of the sample and its transfer is a time-consuming procedure, therefore it was decided to evaluate the fracture toughness of citric acid with Vickers indentation, performed with a

Micromet-1 microhardness tester (Buehler Ltd., Lake Bluff, Illinois, USA). This device could be placed right next to the microscope, and the transfer of the samples to the microscope could be done within a few seconds. Thus, finally fracture toughness values for citric acid could be determined.

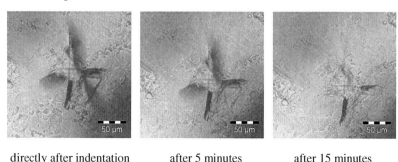

directly after indentation   after 5 minutes   after 15 minutes

Fig. 4.39 Healing of an indent in citric acid monohydrate at ambient conditions

4.4. Identifying the main influences on comminution

4.4.1. Comminution behaviour related to mechanical properties

In order to gain more insight about the interdependency of the measured material properties and breakage parameters, a dimensional analysis was done including f^*_{Mat} and the quantities identified as relevant to breakage by several other authors, as given in Eqs. (2.1) and (2.3) to (2.5) in Section 2.1. The discussion presented here has been published by Meier et al. (2009b). Several simplifications can be made: elastic wave propagation is faster than crack propagation and deformation, therefore both loading and crack propagation can be considered quasistatic, and the influence of v_{el}, v_{fract} and v_d can be neglected (Vogel and Peukert, 2003). Second, since Vogel and Peukert's model considers kinetic energy and particle size explicitly, f^*_{Mat} should be independent of these quantities. The Poisson ratio v is considered to be roughly constant in our experiments and will be neglected in the dimensional analysis as well. The remaining quantities are: f^*_{Mat}, ρ, β_{max}, H, K_C, l_i. The subsequent dimensional analysis with these parameters yields three independent dimensionless numbers.

These are

$$\Pi_1 = \frac{f^*_{Mat}\beta_{max}}{\rho} \; ; \qquad \Pi_2 = \frac{\beta_{max}}{l_i H} \; ; \qquad \Pi_3 = \frac{H\sqrt{l_i}}{K_c} \qquad (4.10)$$

Note that, if E is also included in the dimensional analysis, another dimensionless parameter E/H will be obtained. Although this parameter may also be of importance, it was found to be nearly constant for the examined substances (see Table 4-4), and will therefore be neglected in the following. So, we can derive the following relation:

$$f^*_{Mat} = c\rho\beta_{max}^{y-1}l_i^{z/2-y}H^{z-y}K_c^{-z} \qquad (4.11)$$

where c is a constant; β_{max} and l_i are unknown; as a first approximation, it is assumed that these quantities are similar for the different organic crystalline materials of this study, and therefore they are assumed to be roughly constant. Then, Eq. (4.11) can be simplified as

$$f^*_{Mat} = c_1\rho H^{z-y}K_c^{-z} \qquad (4.12)$$

To obtain the exponents y and z, a 3-dimensional plane fit (using SigmaPlot 9.0) was performed with the quantities $\ln(f^*_{Mat}/\rho)$, $\ln(H)$ and $\ln(K_C)$. This is visualised in Fig. 4.40. The results for the exponents are (z-y) = 2.489 with a standard error of 0.37, and z = 2.874 with a standard error of 0.39, the fit quality was $R^2 = 0.916$. It should be mentioned that also the "critical" substances acetylsalicylic acid and ascorbic acid, which didn't fulfill Eq. (2.24), lie well within the fit plane. Considering the standard errors of the results, it may be assumed that the exponents of H and K_C are equal, with different signs. Indeed, a double-logarithmic fit of f^*_{Mat}/ρ versus H/K_C yields a slope of 2.59 (std. err. = 0.46). For a more practical expression, an exponent of 2.5 should be accurate enough, and thus the following relations are obtained:

$$f^*_{Mat} = c\frac{\rho l_i^{5/4}}{\beta_{max}}\left(\frac{H}{K_c}\right)^{2.5} \quad \text{or} \qquad (4.13)$$

$$f^*_{Mat} = c_1\rho\left(\frac{H}{K_c}\right)^{2.5} \quad \text{with} \qquad c_1 = 2\cdot10^{-13} \; m^{13/4}/J \qquad (4.14)$$

101

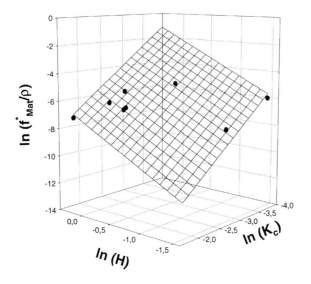

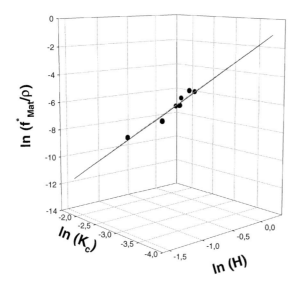

Fig. 4.40 3D plane fit of logarithms of f_{Mat}/ρ, H, and K_c of the examined substances (views from different angles are shown to provide a better visualisation)

This means that f^*_{Mat} is directly correlated with the brittleness index H/K_C, that was already associated with milling behaviour earlier (Taylor et al., 2004a and 2004b). Compound B was not included in this evaluation, because no proper f^*_{Mat} values could be obtained. However, a brittleness index of $23.6 \cdot 10^3 \, m^{-1/2}$ was determined from nanoindentation; according to Eq. (4.14), this would result in an f^*_{Mat} value of 20.7, indicating a very easy breakage of Compound B anyway. The correlation between f^*_{Mat} and the brittleness index is illustrated in Fig. 4.41. It has to be admitted that, looking at the standard errors obtained in data fitting, the exponents of the parameters H and K_C could be varied by about 0.5, and similar correlations containing other exponents would be obtained. However, the correlations given here provide the best fit to the data measured so far.

At this point it is worth taking a look at the unit of f^*_{Mat}/ρ: it is m²/J, which is the reciprocal of a surface related energy. In this regard, f^*_{Mat}/ρ can be interpreted as a measure for the new surface that is created per unit of energy that is put into the system. This is qualitatively correct, since the higher the f^*_{Mat} value, the less energy is needed for comminution of the substance. This behaviour corresponds to either a higher hardness or a lower fracture toughness: the higher the hardness, the more resistance the particle has against plastic deformation. Thus, less energy is consumed for the plastic deformation and more of the input energy remains for the creation of new surfaces, i.e. crack growth and fracture. The same applies to a low fracture toughness: the smaller the resistance against fracture, the less energy is needed to create new surfaces. Accordingly, a substance with a high brittleness index must also have a high f^*_{Mat}.

For an energy balance, the impact energy of the particles can be calculated from the experimental data and compared to the newly created surface area. The surface area can be estimated from the surface weighted mean diameter $d_{3,2}$ from the measurement of the particle size distribution of the comminuted samples by laser light diffraction, coupled with the mass specific energies obtained from the impact velocity. From these calculations, input energies between 10 to 1000 J/m², related to the newly created surface, are obtained. This is very high compared to the surface energies of organic substances which normally lie in the range of several 10 mJ/m². Several factors can explain this huge difference: first, the estimate for the surface area is very rough, since surface roughness is not taken into account which can enhance the real surface area significantly; also, a part of the energy transferred to the particles is converted into plastic deformation. The biggest amount finally is converted into kinetic energy of the particle fragments. Thus only a small part of the energy input is finally converted into surface energy of the new fragments.

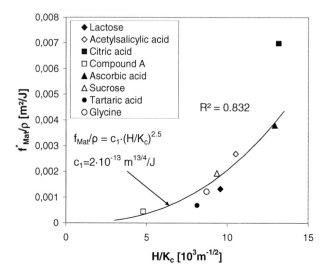

Fig. 4.41 Dependency of f^*_{Mat}/ρ on the brittleness index H/K_C

The same dimensional analysis approach as with f^*_{Mat} can be done for $x \cdot W^*_{m,min}$. Performing the same steps as above, only replacing f^*_{Mat} by $x \cdot W^*_{m,min}$ in the dimensional analysis, we obtain the expression

$$x \cdot W^*_{m,min} = c\frac{1}{\rho}H^{z-y}K_c^{-z} \tag{4.15}$$

The next step in the evaluation of the above correlation is a fitting of the logarithms of H, K_C and $x \cdot W^*_{m,min} \cdot \rho$. The data now scatter much stronger than the f^*_{Mat} values, and therefore only a poor correlation is obtained: $R^2 = 0.627$, $z-y = -3.153$ (standard error = 1.12 and $z = 2.551$ (standard error = 1.19). It was observed from the milling experiments that $x \cdot W^*_{m,min}$ generally has a stronger scattering than f^*_{Mat}. Additionally, the determination of $x \cdot W^*_{m,min}$ becomes increasingly inaccurate for small values of that energy threshold. This might be because of the evaluation procedure, since the results presented are obtained by the population balance method from laser diffraction data. This model has some weaknesses at small breakage probabilities, which in turn affect the result for $x \cdot W^*_{m,min}$ much stronger than for f^*_{Mat} Additionally, stochastically distributed material flaws may have a stronger influence on the milling result in experiments with lower energy input.

Therefore it seems more advisable to obtain the $x \cdot W^*_{m,min}$ value of a substance via its f^*_{Mat} value: already Vogel and Peukert (2003) reported a correlation between f_{Mat} and $x \cdot W_{m,min}$. This correlation is shown in Fig. 4.42, where data from Vogel and Peukert are complemented by data shown in this work. It can be seen that a correlation inarguably exists. Vogel and Peukert's data span over a markedly larger region than the data from this work, and suggest a dependency with a slope of (-1.5) between the logarithms of $x \cdot W^*_{m,min} \cdot \rho$ and f^*_{mat}/ρ. With the exception of citric acid and acetylsalicylic acid, the data from this work also lie near this trendline. As mentioned above, it can be expected that stronger variations in $x \cdot W^*_{m,min}$ are obtained for more brittle substances. Therefore, it seems that the $x \cdot W^*_{m,min}$ values for acetylsalicylic acid and citric acid contain a certain error, which makes it impossible to establish a proper trend when regarding only the substances examined in this work. It is more advisable to derive a correlation from Vogel and Peukert's data, which cover a wider data range. A good correlation ($R^2 = 0.94$) is obtained using

$$xW^*_{m,min} = \frac{1}{\rho}c_2\left(\frac{f^*_{Mat}}{\rho}\right)^{-3/2} \quad \text{with} \quad c_2 = 0.0027\left(\frac{J}{m^2}\right)^{-1/2} \quad (4.16)$$

The presented correlations exhibit a certain scattering. These variations are not only due to the scatter of the measured data, but also some simplifying assumptions were made, and thus some possible influences have been neglected: these are the dimensionless parameter (E/H), the Poisson ratio ν, the fracture energy β_{max} and the initial flaw length l_i, which were assumed to be constant. Considering these parameters in more detail may yield more accurate correlations between breakage parameters and mechanical properties. Also, crystal anisotropy was neglected, which may also influence the results. Nevertheless, obviously the biggest influence on breakage is exerted by the brittleness index, and Eqs. (4.14) and (4.16) provide a means to estimate breakage parameters from indentation data with a sufficient accuracy.

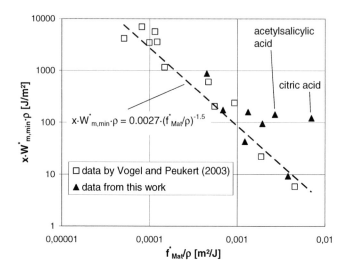

Fig. 4.42 Interdependency of f^*_{Mat} and $x \cdot W^*_{m,min}$

4.4.2. Extension to other material classes

As was already shown by Vogel and Peukert (2003), the Weibull model for the breakage probability applies for different types of material. Also the breakage functions derived in this work seem to be valid for various material classes. However, the correlation between breakage parameters and the brittleness index as derived in this work was established only for crystalline organic solids. It still needs to be checked whether the relation is the same for other material types, such as inorganic solids, amorphous solids, or particles with a compound structure. Therefore, indentation experiments were done with substances whose breakage behaviour was already studied by Vogel and Peukert: those materials are various polymers (two types of polystyrene (PS) and two types of polymethylmethacrylate (PMMA)), two inorganic crystalline solids (ammonium sulphate and potash alum) and two powder coatings (on epoxy and polyester basis). For the PMMA samples, no cracks could be induced at indentation forces up to 1 N, therefore those samples had to be excluded from the evaluation. The experimental results for the other samples are shown in Table 4-6.

In Fig. 4.43, f^*_{Mat}/ρ is plotted vs. the brittleness index for those materials. It can be seen that the inorganic crystals as well as the PS samples coincide quite well with the trendline for the organic crystals. But the powder coatings, especially the epoxy based one, deviate from that trend. As of now, a detailed explanation for this deviation cannot be given; however, it seems likely that the correlation between brittleness and breakage is related to the internal structure of a particle. This explains why inorganic crystals do not differ much from organic crystals in this respect. It may be a coincidence that also the polymers that are only partly crystalline behave in the same way. But the powder coatings have a totally different structure: powder coating consists of a binder, pigments and fillers that form some kind of matrix structure. It seems likely that this structure is the main reason for the observed deviation. The results show that more work needs to be done to get a better understanding of the relationship between a particle's inner structure, its material properties and its breakage behaviour.

Table 4-6 Properties of materials of other types

Sample	f^*_{Mat} [kg/ Jm]	$x \cdot W^*_{m,min}$ [Jm/kg]	E [GPa]	H [GPa]	K_c [MPa$\sqrt{m}$]	BI [10^3 m$^{-1/2}$]
PS 144C	0.13	3.376	3.75	0.179	0.0598	2.99
PS 168N	0.12	5.331	3.71	0.163	0.0705	2.31
Ammonium sulphate	0.97	0.115	24.27	0.464	0.1308	3.55
Potash alum	1.63	0.134	24.04	0.816	0.0989	8.25
Epoxy powder coating	5.39	0.005	6.12	0.415	0.0403	9.63
Polyester powder coating	2.35	0.018	5.12	0.388	0.0526	7.89

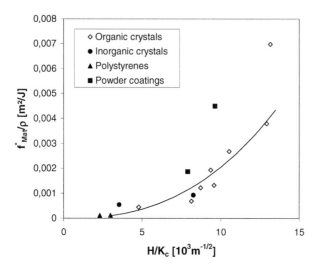

Fig. 4.43 Comparison between organic crystals and other types of material

4.4.3. Relations to crystal structure

In this section, the question shall be addressed whether the comminution behaviour of a substance can be predicted already from its crystal structure. From the findings shown above, this should be possible if hardness and fracture toughness can be predicted correctly. Computer programs are available that use molecular dynamics to model crystal morphologies and calculate a wide variety of mechanical properties such as hardness or elastic constants. Examples for such programs are DMol³, that uses density functional theory (DFT), or GULP, which uses forcefield methods. Even if it is not yet possible to get values for the fracture toughness directly from those programs, a close look at the crystal structure can be useful. To obtain a first estimate for the toughness of a material, it may help to identify the slip systems of a crystal and determine their frictional stress. A slip system consists of a crystallographic plane and a direction in which slip occurs. As a general rule, the preferred slip planes in a crystal are the most dense packed ones, as they have the greatest separation distance between the single layers; preferred slip directions are also the most close-packed directions, for which the atomic slip distance is a minimum (Courtney, 2000). The slip systems depend on the geometry of the crystal structure and on the specific interactions between the atoms or molecules of the crystal. For metals, which usually consist of single atoms and of relatively simple crystal structures (face-centered cubic, body-centered cubic and hexagonal close-packed), the slip systems are well known. However, pharmaceutical compounds often possess more complicated crystal structures, such as monoclinic or triclinic. Also, inter-molecular interactions (e.g. polar interactions and hydrogen bondings) may influence the slipping. This makes it more difficult to identify the slip systems in these crystals. Nevertheless, a detailed look at the crystal structure with the help of molecular modeling may help to judge the mechanical behaviour of a crystal.

This can be illustrated with the example of Compound A: we may assume that breakage is constricted if crystal planes can slide very easily on each other, i.e. if plasticity and toughness are enhanced. It seems that this is what happens in Compound A: Fig. 4.44 shows crystals of Compound A after a very simple initial test - the particles were hit manually by a hammer in order to get a visual impression of their mechanical behaviour. As can be seen, the crystals possess a layered structure; the energy loaded onto the crystals was consumed mainly not for breakage but for plastic deformation, as can be seen well on the right hand side of Fig. 4.44: the image reminds to the deformation that takes place when a telephone book is bent and the pages slide past each other.

Another interesting question is whether the comminution behaviour can be estimated already from the molecular structure of a substance. This surely is not an exhaustive consideration, but in the case of Compound A, a hint for flexibility may be seen in its structure. A simplified sketch of the structure of Compound A is shown in Fig. 4.45. In this sketch, some side groups are omitted; heteroatoms are not labelled as such, also double bondings are not shown. However, the main feature of the structure that will be discussed in the following is still clearly visible: one single ring (on the left hand side of the sketch) that is connected to the rest of the molecule by a chain of two atoms.

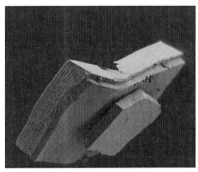

Fig. 4.44 Crystals of Compound A after hitting with a hammer

A large part of this molecule is relatively stiff, and only minor changes in its spatial orientation can be achieved by configurational changes, e. g. rotations around a binding. However, if a rotation is done around the axes of the two-atoms-chain, a major geometric change happens as the single ring on the left hand side is moved around: its orientation relative to the rest of the molecule will be changed significantly, as can be seen in Fig. 4.45. Regarding a single molecule, this rotation is not sterically hindered and can be done very easily.

Although this is just an assumption now, it may be possible that similar effects happen if the molecule is bound within a crystal: when a stress is applied to a crystal of Compound A, the molecules possess a certain orientational freedom and may move their single ring as a response to the imposed stress. This does not necessarily lead to a new crystal structure, but to a certain local disorder of the present structure. In this way, the plasticity of the crystal may be enhanced, and the flexibility of the single molecules might result also in enhanced flexibility of the crystal structure, or in other words: local defects may be induced quite easily in such a crystal. This observation becomes particularly interesting when considering another substance: Taylor et al. (2004a) examined several pharmaceutical compounds by nanoindentation. The by far least brittle substance of their study was voriconazol - a compound that has similar features as Compound A: two heterocyclic groups can be rotated in a similar way, thus enabling the molecule to change its shape. Probably, the same effects as explained for Compound A can happen also within crystals of voriconazol.

Since only two substances are regarded now, and more information about the crystal structure and other properties of voriconazol is not available to the author, these considerations surely are incomplete. Still, if more substances with similar poor comminution behaviour are identified, these observations may be used as a starting point to find out more about the influences of crystal and molecular structure on mechanical properties.

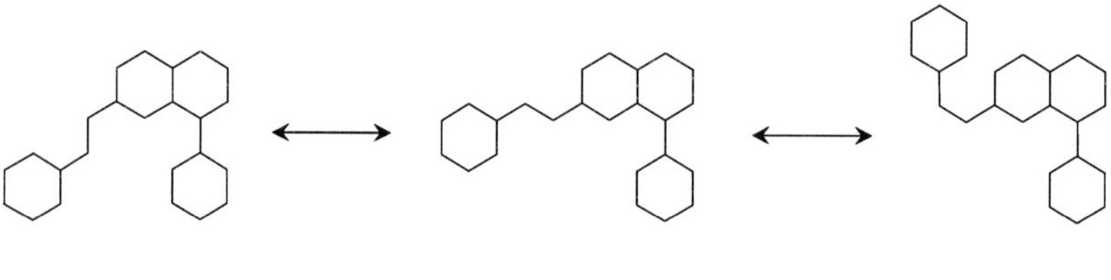

Fig. 4.45 Conformational changes of Compound A; note that the shown structure is simplified - heteroatoms, side groups and double bondings are not displayed

110

4.5. Tabletting

Four substances were tested in tabletting experiments, using both the "in-die" and the "out-of-die" evaluation method. Samples of various particle sizes were used for "in-die" experiments, as they were obtained from the initial milling experiments. Because of the greater experimental effort of "out-of-die" measurements, only one sample of each substance was taken - samples milled at 6 bar in the spiral jet mill (4 bar for acetylsalicylic acid) were used. The Heckel plots are shown in Fig. 4.46. The yield strength values obtained from these plots as well as the elasticity parameter K_{et} are given in Table 4-7. As can be seen from Fig. 4.46, the shape of the curves of acetylsalicylic acid and Compound A is strongly dependent on the particle size: the finer the powder sample, the smaller the slope of the Heckel curve. The curves of lactose and Compound B, however, show no dependency on particle size.

A closer look shall be taken at the correlation between the parameters derived from tabletting (yield strength σ_0 and elasticity parameter K_{et}) and hardness and Young's modulus. Recall that $P_y = 3\sigma_0$. For three of the substances, E/P_y is around 100. According to Eq. (2.34), $H/P_y = 3$ should be obtained for these substances. This is the case, as can be seen from Fig. 4.47a: the data points lie close to the line $P_y = H/3$. The substance having the largest deviation from this line, acetylsalicylic acid, has an E/P_y of only 50. According to Eq. (2.35), $H/P_y = 1.93$ should be obtained. Again, the measured value of 2.03 corresponds quite well to the theory.

Table 4-7 Yield strength σ_0 and elasticity-parameter K_{et} obtained from Heckel plots

	σ_0 (in-die) [MPa]	σ_0 (out-of-die) [MPa]	K_{et} [MPa]	E/P_y [-]	H/P_y [-]
α-lactose monohy.	42.15	86.08	228.6	83.0	3.36
Acetylsalicylic acid	5.12 ..13.37	36.49	63.3	49.7	2.03
Compound A	6.41 ..11.93	31.56	57.6	104.9	2.62
Compound B	16.86	34.61	95.6	103.1	3.27

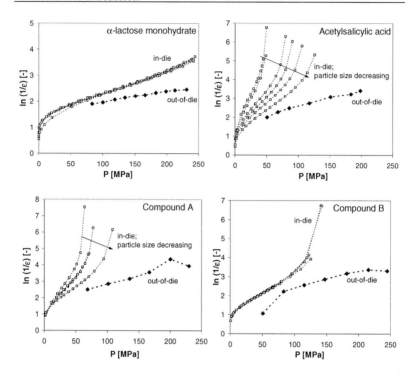

Fig. 4.46 Heckel plots of the four tested substances

The parameter K_{et} is compared to the Young modulus in Fig. 4.47b. As far as can be judged from the few data points, a linear correlation seems to exist. This is not surprising since K_{et} is derived from the differences between in-die-measurements (where the elastic deformation is still present in the material) and out-of-die-measurements (where elastic deformation is removed). Therefore, K_{et} represents -like the Young modulus- a measure for elastic deformation.

These observations show that it should be possible to determine hardness and Young's modulus also from powder compression data. However, the correlation between K_{et} and E needs to be examined in more detail, on the basis of many more substances. This approach was not followed further, because it was beyond the focus of this work. For determination of hardness and Young's modulus, nanoindentation has proven to be a fast and powerful tool that additionally can be used also for the determination of fracture toughness.

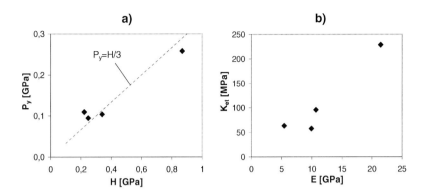

Fig. 4.47 Correlation between parameters obtained from Heckel analysis and mechanical properties (σ_0 values from out-of-die measurements)

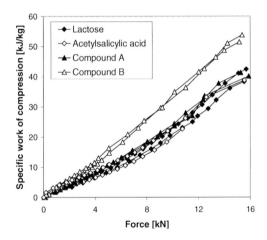

Fig. 4.48 Specific work of compression obtained from tabletting experiments

Fig. 4.48 shows the specific work of compression over the applied load. For each substance, the work of two samples with different size distributions are shown. For three of the substances, the specific work increases in a similar way; only for Compound B a noticeable higher work is applied during compression. This corresponds to the poor flowability of this substance that is observed also in shear experiments (see Chapter 4.6). Unfortunately, the energy released during unloading could not be measured with the testing machine.

4.6. Shear tests

Shear tests were conducted in order to characterise the flowability of feed material and comminuted powders and the friction on different wall materials. A direct comparison of the results for the feed materials is of little use, since the particle sizes are quite different. Therefore, milled samples that possess roughly the same median particle size ($\sim$ 15 µm) were used for comparison. Samples of lactose and Compound B were taken from milling trials in the Escolab spiral jet mill, while the acetylsalicylic acid samples were milled in the Condux pin disk mill. The general trends - the bigger the particle size, the better the flowability and the less the wall friction - were observed with all the powders. The milled samples examined here possess similar flowabilities around 2 (see Table 4-8). This value is generally considered as the border between cohesive and very cohesive behaviour. Compound B has a slightly poorer flowability than lactose and ASA. A complete overview over the results of all measurements, including tests at different consolidation stresses, is given in Appendix E.

For assessing the wall friction, three different wall samples are used: one steel plate with a roughness of $R_a = 0.97$ µm (referred to as "rough steel"), one steel plate with $R_a = 0.21$ µm ("smooth steel"), and a PTFE plate ($R_a = 1.16$ µm). The results are depicted in Fig. 4.49, the wall friction angles are given in Table 4-8. The shear stress rises linearly with the normal stress. With PTFE, the first measurement point done with a freshly prepared powder bed always possesses a higher friction, which can not be reproduced when the experiment is repeated with the same bed. Obviously, slipping becomes easier once the first resistance (in the first shearing) is overcome.

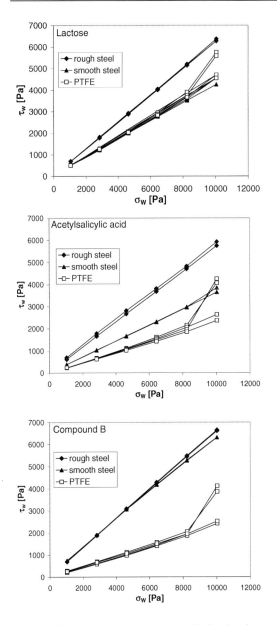

Fig. 4.49 Wall shear stresses measured with the ring shear tester

Table 4-8 Wall friction angle φ and flowability ff_c of the tested powders

	x_{50}	φ on rough steel	φ on smooth steel	φ on PTFE	ff_c at $\sigma_1 \approx 10000$ Pa
	[μm]	[°]	[°]	[°]	[-]
Milled samples					
Lactose	12	32.1	23.7	24.1	2.0
Acetylsalicylic acid	18	29.5	19.7	15.5	2.0
Compound B	15	33.4	31.7	13.5	1.8

Big differences can be seen regarding the different samples on different wall materials: with rough steel, the wall friction is similar for all three tested substances (φ ≈ 30°), Compound B having the highest friction. The smoother steel surface reduces the wall friction of lactose and acetylsalicylic acid, but for Compound B, the friction stays almost the same. On PTFE, the friction is further reduced for ASA, but not for lactose. For Compound B, friction is now reduced significantly. These results show the remarkable influence that the choice of wall material has. It is interesting to note that the roughness of the steel sample has only little influence on the wall friction of Compound B. The measurements give a first hint to a more sticky behaviour of Compound B, but they do not show a clear distinction.

4.7. Atomic Force Microscopy

This section contains the results of adhesion measurements done by Atomic Force Microscopy. First, experiments were done with spherical reference particles (gold and glass). Later, interactions between drug particles (irregularly shaped) and drug substrates are assessed.

4.7.1. Gold spheres on lactose

Surface scans of different lactose samples that were used as substrates are shown in Fig. 4.50. Measured crystal faces were (010) and (110). The crystal face could be determined from the shape of the crystal. Spherical gold particles for these measurements were made in a flame reactor that is described in detail by Götzinger (2005). Adhesion force distributions of a gold sphere with a diameter of 5.5 μm in contact with these surfaces are depicted in Fig. 4.51. It can be seen that different faces exhibit similar force distributions. In several cases, bimodal force distributions are observed (L6 and L8), this obviously owing to some very fine particles adhered to the surfaces, as can be seen on the images in Fig. 4.50. These fines stem from the crystallisation process.

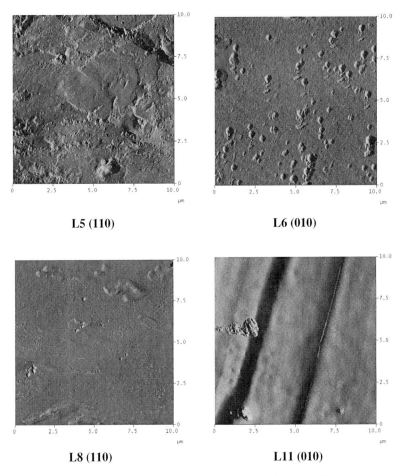

Fig. 4.50 Surface scans of lactose samples: L5-L8: prepared in fluidised bed crystalliser; L11: cleaved face

An exception is the sample "L5". The crystal comes from the same crystallisation batch as L6 and L8, but obviously, an improper washing or drying happened to that crystal. Here, adhesion forces are significantly lower, and the force distribution curve has a different shape, which rather looks like a Weibull distribution function. This type of distribution can appear on rough surfaces (Götzinger and Peukert, 2004). In fact, the surface of this sample has a markedly higher roughness than the other samples (rms = 0.41 nm as opposed to around 0.2 nm for the other samples; see Table 4-9). Obviously, the sphere cannot make a proper contact with the sample surface due to the irregular shape of the surface in most cases. From measurements with larger gold spheres, force distribution curves of the same shape but with higher force values are obtained. From the adhesion theory of Hamaker (1937), a linear dependence of the adhesion force from the particle radius is expected. The measured data coincide with this theory, as can be seen from Fig. 4.52, where F_{90}-values of the force distributions have been plotted over the sphere diameter. F_{90} was used instead of F_{50} for this comparison of different sphere radii, because this value is less affected by the bimodalities due to surface roughness which is observed mainly with the larger spheres.

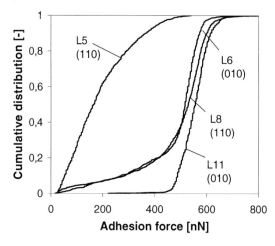

Fig. 4.51 Adhesion force distributions of a 5.5 μm gold sphere on different lactose samples

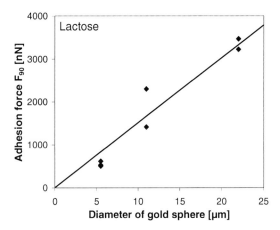

Fig. 4.52 Adhesion forces of gold spheres with different diameters on lactose

4.7.2. Van-der-Waals forces of other substances

Surfaces of the other substances studied are depicted in Fig. 4.53. It should be noted that it was hard to prepare crystals of Compound A without any fines adhering to the surface after crystallisation, therefore all measured force distribution curves are bimodal, the upper part of the curves representing the forces obtained by proper contacts of the sphere with the surface.

Fig. 4.54 depicts representative adhesion force distribution curves of the four different crystalline materials in contact with the same 5.5 μm gold sphere. Fig. 4.55 shows F_{90} values of measurements with spheres of different size on acetylsalicylic acid and Compound B. With Compound A, only the 5.5 μm gold sphere could be used for measurements, because the adhering fines prohibited proper sphere-substrate contacts with larger spheres. As can be seen from these figures, the mean adhesion forces of all substances are similar. Compound B, the substance that could be expected to have the highest adhesion force, shows even the lowest van-der-Waals interaction of all substances.

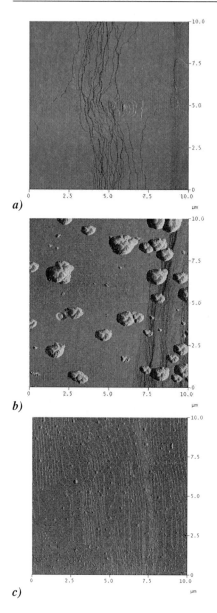

Fig. 4.53 a) Surface scan of cleaved ASA crystal, (001)-face; b) Compound A; c) Compound B

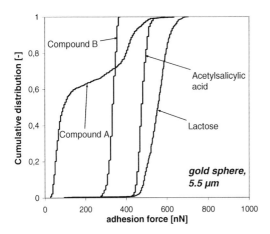

Fig. 4.54 Adhesion force distribution curves of a gold sphere (5.5 μm diameter) on different substances

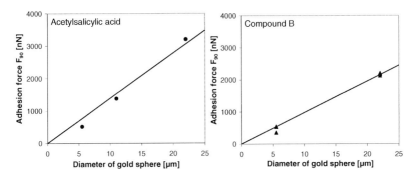

Fig. 4.55 Adhesion forces (F_{90}-values) of gold spheres of different sizes on acetylsalicylic acid and Compound B

The models by Hamaker (1937) and by Rabinovich et al. (2000) were employed in order to compare the measured data with the theory. Hamaker constants of the solids were estimated from the dispersive components of the surface energy which was measured by inverse gas chromatography (see Chapter 4.8), via the relation

$$A = 24\pi z_0^2 \gamma_s^d \tag{4.17}$$

(Israelachvili, 1992), where z_0 is the minimum contact distance, γ_s^d is the dispersive component of the surface energy. The results of the adhesion force

121

calculations are given in Table 4-9. In this case, the median values F_{50} are given for the measured adhesion forces, because these are more directly comparable with the values derived from the models than F_{90}. (F_{90} is the more suited quantity for comparison of different sphere sizes, as with larger spheres, bimodalities may influence the F_{50} value.) With the Hamaker model, much higher adhesion forces are predicted than were actually measured. The Rabinovich model takes the surface roughness of the samples into account and yields adhesion forces that lie within the range of the forces measured. A more precise prediction of adhesion forces is not possible with this relatively simple model, because it does not take into account other influences such as deformations or local variations of the surface roughness.

Table 4-9 rms roughness values, calculated and measured adhesion forces of a 5.5 μm gold sphere (Hamaker constants were calculated from iGC data)

Substance	Sample	rms-roughness [nm]	Hamaker constant $[10^{-20}J]$	F_{adh} (Hamaker model) [nN]	F_{adh} (Rabinovich modell) [nN]	F_{50} (experimental) [nN]
Lactose	L5	0.410	9.1	3206	146.5	185.6
Lactose	L6	0.207	9.1	3206	393.3	516.2
Lactose	L8	0.224	9.1	3206	354.5	528.9
Lactose	L11	0.124	9.1	3206	718.9	556.1
ASA	1	0.121	7.5	2906	668.5	474.1
Comp. A	1	0.253	9.2	3225	302.2	241.6
Comp. A	2	0.222	9.2	3225	360.8	408.5
Comp. B	1	0.130	8.5	3097	660.3	597.2
Comp. B	2	0.295	8.5	3097	233.5	330.9

4.7.3. Measurements with glass spheres

Glass spheres were chosen as a polar adhesion partner. In the first experiments, glass spheres functionalised with aminopropyltriethoxysilane (obtained from Potters Europe GmbH, Kirchheimbolanden, Germany) were used. These were chosen to obtain a well-defined polarity on the surface. But these spheres turned out to be very rough, the obtained adhesion force distributions were very broad and scattering. In this case, adhesion was dominated by the roughness of the functionalised spheres. Therefore, untreated glass spheres were used instead. Now, better measurements were possible: adhesion forces were larger and less scattering. The same measurement, repeated with the same sphere on the same crystal, gave reproducible results. However, if experiments were repeated with another glass sphere on the same crystals, the results varied a lot, even although the spheres were all conditioned in the same way, i. e. heating overnight at 150°C. An example is shown in Fig. 4.56. Measurements with a 10 μm glass sphere show strong adhesion of Compound B, while the sphere adheres far less strong to ASA than to the other substances; in another measurement with a 7 μm sphere, totally different forces are obtained, the adhesion on Compound B is even less than on ASA this time. Since the single measurements were reproducible, the reason for these findings seems to lie in the different glass spheres: the most probable explanation is that the polarity of a glass sphere is all but well-defined. Its spatial distribution on the surface is very inhomogeneous. Therefore, every sphere will exhibit a different polarity at the point of contact with the substrate, resulting in very different results as shown in Fig. 4.56. Thus, glass spheres turned out to be too undefined to serve as a reference system. For a proper assessment of polar interactions, a smooth, well-defined adhesion partner with well-defined homogeneous polarity on the surface would have been necessary.

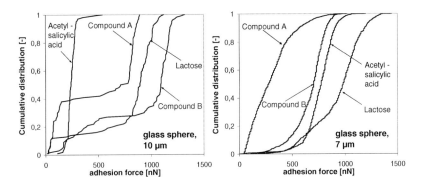

Fig. 4.56 Adhesion force distributions of a 10 μm glass sphere (left) and a 7 μm glass sphere (right) on different substances

4.7.4. Adhesion at elevated levels of relative humidity

The influence of humidity on the adhesion force was examined. Measurements were performed at 0, 20, 40, 60 and 80 % relative humidity. Before each measurement, the cell was flushed for 45 minutes with a nitrogen stream of the respective humidity. As reference particles, gold spheres, a glass sphere, and drug particles were used on the different substrates. For the measurement of one force distribution, 100 single force measurements were done on a lattice of 10x10 points at 1 µm distance. For each force distribution, the F_{50} value was taken, and the adhesion force relative to the F_{50} force measured at 0 % r.h. was calculated. The resulting curves are displayed in Fig. 4.57. These findings have to be interpreted together with the results of dynamic vapour sorption experiments that are shown in Fig. 4.58. (DVS measurements were done by Phillippe Piechon, Novartis Pharma, Basle, Switzerland, using a DVS system by Surface Measurement Systems Ltd., Alperton, UK). As can be seen, three of the model substances show practically the same adsorption behaviour. For these three substances, the adsorbed amount of water at 80 % r.h. corresponds roughly to a monolayer, i.e. adsorption is not very strong. Only Compound B adsorbs some more water than the other substances.

In contact with the gold sphere, no big changes of adhesion occur with all the substances. Obviously, adsorption is quite low also on the gold sphere, and the adsorbed amount is too low to influence adhesion noticeably. With the glass sphere, all the adhesion forces rise by a factor of about 2. Here, adhesion on the glass sphere seems to be much stronger than on the organic crystals, therefore the adsorption on the sphere dominates the adhesion. These findings are in accordance to those of Jones et al. (2002), who studied interactions between glass spheres and hydrophobic and hydrophilic glass and silicon surfaces: generally, they found that adhesion increases with relative humidity for the hydrophilic surfaces, while no change was observed with hydrophobic surfaces. The same seems to apply for the experiments shown here with gold and glass.

Measuring the change of self-adhesion of compound particles on a substrate of the same material, the forces change only slightly for lactose and Compound A. For Compound B, the adhesion force increases at higher humidities, which may be attributed to the somewhat higher adsorption. In the case of acetylsalicylic acid, a very strong increase of the adhesion force during one measurement is observed. This increase is not reversible, i.e. when another measurement at 0 % r.h. is done after the experiment, the adhesion forces did not return to the initial level again. From these observations it seems that the ASA particle is damaged or changes its shape under the influence of the humidity in combination with the mechanical stress imposed during the measurement; the increase of adhesion seems attributed to these effects rather than to the humidity itself.

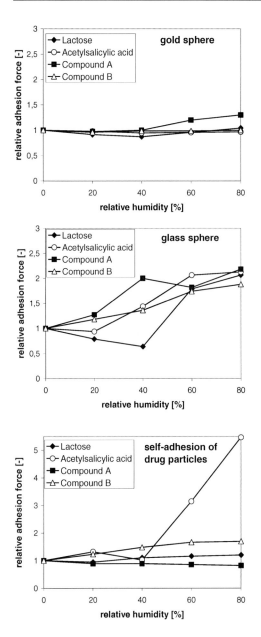

Fig. 4.57 Relative adhesion forces at different humidites in different systems

125

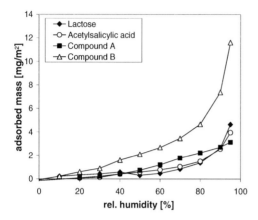

Fig. 4.58 Adsorption of water on the organic substances (surface measured by BET)

4.7.5. Adhesion on wall substrates

Adhesion forces of drug particles were measured on drug substrates and wall materials. Drug particles for these measurements were chosen in the size range between 5-15 µm, found in non-milled material or material milled in the pin disk mill. Material milled in the spiral jet mill was not used in order to avoid effects of possibly amorphous surfaces. These particles possess an irregular shape, and the knowledge of the exact adhesion force is of no use, since the contact geometry is unknown. Therefore adhesion forces can only be compared relative to each other. Each AFM tip that carried a particle was measured on a steel substrate, a PTFE substrate and on a crystal of the same substance as the particle. The relative adhesion force is the adhesion force of a particle on a wall material, compared to the adhesion force on a crystal of the same material:

$$F_{rel} = \frac{F_{wall}}{F_{self}} \qquad\qquad (4.18)$$

Measurements were performed with several particles of each substance, glued to tipless AFM cantilevers. On each substrate, a force distribution curve consisting of 100 single measured forces was obtained. The median value of each distribution was used for evaluation. In fact, the jump-out-distance Δz from the point of zero force (particle in contact) to the jump-out is a sufficient measure for the relative adhesion force of a particle on different substrates. Thus, determination of spring constants could be spared. As substrates were chosen: steel, PTFE, and flat crystal faces of the four model substances.

Table 4-10 Relative adhesion forces of the drug particles on different substrates

Particle material	F_{rel} on steel substrate	F_{rel} on PTFE substrate
α-lactose monohydrate	0.11	0.56
Acetylsalicylic acid	0.46	2.69
Compound A	0.41	0.58
Compound B	0.37	0.35

The results of these measurements are given in Table 4-10. The first important outcome of these measurements is that all values except one are smaller than unity, i.e. the intrinsic self-adhesion within a powder is higher than the adhesion on a wall material. This may lead to the conclusion that, if sticking is to be avoided, the particle-wall contact is the decisive point and not the cohesiveness of the powder itself. But in reality, often a thin powder film is found on equipment walls (also on the wall samples in the shear experiments), suggesting that adhesion on the walls is higher than within the powder. However, one has to keep in mind that in the case of a particle-particle contact, a different geometry applies than for a particle-wall contact. Here, we compared only this latter case. So, even if the intrinsic adhesion of a drug particle to a crystal of its own kind is higher than to a wall, it still may be more likely to stick to a wall than to another particle due to geometric reasons.

The one exception mentioned above is acetylsalicylic acid, where the adhesion force on PTFE is higher than on ASA itself. At this point it can only be speculated about the reasons, but it seems likely that the problems mentioned earlier also cause trouble when measuring with ASA particles on the AFM tip: since ASA breaks very easily, there is always the danger that a part of a particle breaks away during the measurements, thus changing the contact geometry and therefore also changing the adhesion forces. Second, ASA possesses a quite high solubility in almost every solvent. If an ASA particle was glued to a cantilever, very little glue had to be used, otherwise the glue would dissolve the particle after some time. It was also observed that adhesion forces of ASA particles rose constantly during measurements, as if the glue worked as a plasticiser in the crystal. Several times, ASA particles fell off the cantilever during the measurements. Therefore, the results for ASA particles should be regarded with caution.

The second interesting aspect that can be seen from Table 4-10 is that the relative adhesion force on PTFE always is greater than on steel. This is not what one would naturally expect; also the shear tests (see Chapter 4.6) suggest just the opposite. To explain this result, one has to keep in mind that the shear experiments characterise a kinematic friction, while the AFM experiments do not detect this at all: the AFM cantilever is pulled straight upwards, therefore kinematic friction cannot have any influence. This again agrees with the observations made in the shear tests with PTFE (compare to Fig. 4.49): with all powders, a higher shear stress was measured in the first experiment with a new bed; only after a first resistance is overcome, the markedly smaller kinematic friction is observed. Secondly, also local deformations in the contact zone may have an influence: PTFE has a markedly smaller hardness and elastic modulus than the organic crystals examined here (hardness: 0.05 GPa compared to > 0.2 GPa; Young's modulus: 1.3 GPa compared to > 5 GPa; data for PTFE are taken from McCook et al., 2005). Therefore, the deformation was presumably much higher in the PTFE experiments than in those using other substrates, which also results in a higher adhesion force.

Apart from these findings, again no clue was found for an exceptional sticky behaviour of Compound B. This behaviour becomes apparent during milling. Obviously, the conditions leading to such a behaviour could not be reproduced in the AFM experiments. In the mill, very high load levels are reached during an impact, while in the AFM, the force that can be applied to a particle-substrate contact is limited by the flexibility of the cantilevers: with a maximum spring constant of 6 N/m and a maximum z distance of 6 μm, a load of 36 μN could be obtained. At these "low-load" measurements, no intrinsic stickiness was detected for Compound B. Also in the shear tests, that may be seen as "medium-load" experiments, only a slight trend towards poor flowability became visible. Therefore, it seems likely that the sticky behaviour is rather a geometry problem: at higher loads, deformation of the particles is increased. Thus, the contact area is increased, resulting in a better adhesion. Also a smooth surface, without any fine particles adhered to, enhances the contact between two surfaces. As can be seen from the SEM images (see Fig. 4.7 to Fig. 4.10), Compound B has very smooth surfaces, also after milling; breakage happens preferrably perpendicular to the long axis, and the long smooth sides of the crystals stay smooth as they are and can form good contact with other smooth particles or a smooth wall. From the measurements shown in this work, it has to be concluded that the scaling of Compound B is supposedly due to these geometric reasons rather than to intrinsic material properties. As a consequence, such a problematic behaviour could possibly be avoided by the generation of particles with a controlled shape or surface roughness, which could be achieved e.g. by choosing appropriate crystallisation or drying conditions.

4.7.6. Adhesion between different organic materials

Adhesion forces were measured also between the four model substances, i. e. a lactose particle on a Compound A substrate and so on. As described before with the wall adhesion measurements, only the relative adhesion forces can be compared. The mean values of the obtained relative adhesion forces are given in Table 4-11.

Table 4-11 Adhesion forces of particles on different substrates relative to their self-adhesion forces

Particle	On substrate:			
	Lactose (010)	ASA	Compound A	Compound B
Lactose	1.00	1.18	0.73	1.21
Acetylsalicylic Acid	0.80	1.00	1.00	2.69
Compound A	1.43	0.68	1.00	0.26
Compound B	1.18	0.80	0.64	1.00

It should be expected that the relative adhesion force e. g. of a lactose particle on an ASA substrate is the reciprocal of the relative adhesion force of an ASA particle on a lactose substrate. In fact, this holds true when lactose and ASA are compared directly, and also for lactose and Compound A, as can be seen from the values in Table 4-11. But this does not work comparing ASA with Compound A, and not for any comparison with Compound B. It is likely that polar interactions between different molecular groups on the surfaces play a decisive role, therefore it can be important which crystal face and which part of an irregularly shaped particle come into contact.

This is underlined by measurements on two different lactose faces: particle adhesion was measured both on a (010) face and on a (100) face. Table 4-12 shows the measured adhesion forces of different particles on both of these lactose substrates. As can be seen, the forces on the different substrates differ, depending on the kind of particle. Adhesion of organic particles is always stronger on the (010) face than on the (100) face. The gold sphere adheres to both faces with practically the same force. In the case of gold, only van-der-Waals interaction takes place, so the differences in the other experiments may be explained by polar interactions, induced by different molecular groups.

Table 4-12 Adhesion forces of different particles on different lactose samples

Particle	Δz on Lactose (010) [nm]	Δz on Lactose (100) [nm]	Ratio of adhesion forces $F_{(100)}/F_{(010)}$ [-]
Lactose 1	71.4	54.6	0.77
Lactose 2	177.7	123.0	0.69
Lactose 3	1845.5	1677.7	0.91
ASA 2	486.9	279.0	0.57
ASA 3	40.1	25.0	0.62
Compound A 1	3434.4	3217.3	0.94
Compound A 2	480.5	424.5	0.88
Compound B 1	1398.4	1052.2	0.75
Compound B 2	102.5	51.7	0.50
Compound B 3	374.1	182.4	0.49
Gold, 5.5 μm	550.8	530.3	0.96

From these observations, the conclusion can be drawn that particle adhesion depends strongly on the local surface properties of both adhesion partners and their interaction. Therefore, it is not possible to create a general ranking of different materials ("what adheres better to what?"), rather, the interaction of both adhesion partners of interest has to be characterised directly.

4.8. Inverse Gas Chromatography

Adsorption on the pharmaceutical powders was assessed by use of inverse gas chromatography. Henry coefficients of adsorption were calculated from the measured retention times. Additionally, the dispersive component of the surface energy was determined. Both samples before milling and samples milled in the Escolab spiral jet mill and the Condux pin mill were studied. Table 4-13 shows the average values of dispersive surface energies. Henry coefficients of alkanes are visualised in Fig. 4.59. The numerical values of the Henry coefficients are given in Appendix F. For lactose, literature data are available for the dispersive component of the surface energy. Newell et al. (2001) determined surface energies of 31.2 mJ/m² for crystalline lactose, and 37.1 and 41.6 mJ/m² for amorphous spray-dried lactose and for milled lactose, respectively. Ticehurst et al. (1996) measured values between 40 to 44 mJ/m² for different batches of crytalline lactose. Planinsek et al. (2003) measured also values between 41 and

44 mJ/m² for crystalline lactose under varying conditions (at different flow rates, column lengths and column materials). The data presented in this work lie at the upper boundary of that range. The Ticehurst data show that batch-to-batch variations can have a noticeable influence on the results from iGC: although different batches are indistinguishable by other methods such as Raman spectroscopy, X-ray powder diffraction or DSC, the surface properties may be altered by minor changes in surface crystallinity or purity. This may also be caused by different processing steps done by the manufacturer, e. g. a sifting of the powder may render the powder surface amorphous, while the amount of amorphous material remains too small to be detected by the other methods. The cited references obtained their lactose samples from different manufacturers; the lactose used in this work comes from the same supplier as the samples of Planinsek et al. (DMV International). Also the fact that the surface energy in the experiments shown here stays the same for the lactose before milling and after commminution in a spiral jet mill at high pressure (see Table 4-13) supports the thesis that the reported variations are mainly due to different processing conditions.

Table 4-13 Dispersive components of surface energy at 30°C

Substance		γ^d [J/m²]
Lactose	before milling	44.4
Lactose	spiral jet mill, 6 bar	44.2
Acetylsalicylic acid	pin mill, 8000 rpm	32.7
Acetylsalicylic acid	pin mill, 19000 rpm	35.4
Acetylsalicylic acid	spiral jet mill, 2 bar	37.4
Acetylsalicylic acid	spiral jet mill, 6 bar	35.8
Compound A (Batch 01.01<250)	before milling	44.7
Compound A (Batch 98 905 MIC)	spiral jet mill	53.5
Compound B	before milling	41.7
Compound B	spiral jet mill, 2 bar	41.2
Compound B	spiral jet mill, 6 bar	41.1

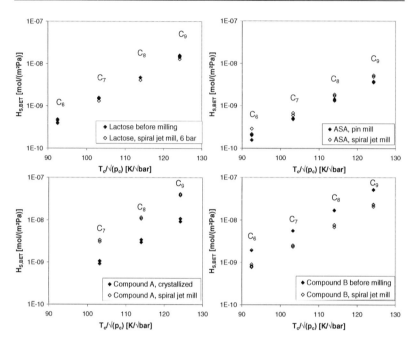

Fig. 4.59 Henry coefficients of the examined solvents alkanes on different substances (closed symbols = samples before milling (for ASA: samples milled in pin mill, low energy input); open symbols = samples milled in spiral jet mill, i. e. high energy input)

The data obtained with hexane as molecular probe show a bigger scattering than data obtained with the other alkanes. This is due to the small retention times of hexane that are more susceptible to errors. With lactose, as already mentioned, both the dispersive surface energy and the Henry coefficients stay practically the same both for feed material and milled powder. In the case of acetylsalicylic acid, a slight increase of surface energy and Henry coefficients is observed, when the powder is milled. A difference between the samples milled in the pin mill and the samples milled in the spiral jet mill is visible. The feed material of ASA was too coarse to be measured with iGC: the amount of powder that could be put into a column had only a small surface area, and the obtained retention times were too short for a proper evaluation. Interestingly, for Compound B, the milled samples show the same Henry coefficients and surface energy, while the sample before milling possesses even higher Henry coefficients and a slightly higher surface energy. As of now, this behaviour - surface energy and Henry coefficients decreasing upon milling - can not be explained. The most significant difference between feed material and comminuted powder was observed with Compound A: not only are Henry coefficients of the micronised samples significantly higher, also the surface energy is about 20 % higher. Different batches were used for these experiments, since the originally crystallised and milled batch of Compound A was again too coarse for iGC measurements. Also other measurement problems appeared with Compound A: the chromatograms of the micronised samples showed a strong tailing, and the measured retention times were dependent on the concentration of the probe molecules. The tailing is a sign that the gas-solid adsorption equilibrium is not in the Henry region anymore, i. e. the concentration of the probe molecules was too high. Only with a very high dilution, tailing was finally reduced, and reproducible results could be obtained. As already mentioned, the results for Compound A are significantly different for samples before and after milling, while for the other substances, differences are relatively small. The strong increase of surface energy and of the Henry coefficient could be a hint for a high breakage energy necessary to create new surfaces of Compound A particles. However, a more detailed study of more substances are necessary to come to a definite conclusion, whether the change of surface properties upon milling can serve as a measure for the grinding performance of a substance.

5. CONCLUSIONS AND OUTLOOK

In this work, the milling behaviour of organic crystalline solids was studied. These materials were chosen to serve as model substances for pharmaceutical powders. Two of the substances, Compound A and Compound B, were known before to possess a critical behaviour upon milling: while it is very hard to comminute Compound A even at very high power input, Compound B clogs the mill very quickly, before noteworthy comminution is achieved. Starting from these observations, the experiments in this work cover two different aspects: the breakage behaviour of the materials themselves, and the transport properties of the powders, which are influenced by particle interactions. Mechanical and breakage properties of the substances were assessed by single particle comminution experiments and by nanoindentation. Surface properties and particle interactions of the powders were studied by shear tests, atomic force microscopy and inverse gas chromatography.

A device was presented to perform single particle impact experiments with fine powders (~25-500 µm). It could be shown that this device is capable of producing reliable results in this size range. In order to allow easier testing of finer powders and to reduce the amount of material required, a new method was developed to determine breakage probabilites from size distributions measured by laser light diffraction. The results obtained by this method are well comparable with the results of the sieve analysis. It could be shown that the concept of describing the milling performance by means of a material function can be applied also to pharmaceutical powders. Moreover, it became obvious that Compound A indeed requires much high energies for comminution than the other substances tested. It could be shown that single particle impact testing allows the characterisation of grinding behaviour of pharmaceutical materials using much lesser amounts of substance than the "standard" milling trials, thus reducing development costs.

The breakage functions of the studied materials could be modeled by bimodal logarithmic normal distributions. The variation of the parameters of this model with impact energy was studied. It was found that the breakage functions can be described by a universal set of equations, valid for all substances studied, also for substances of different material classes. The same parameters f^*_{Mat} and $x \cdot W^*_{m,min}$ that describe the breakage probability of a material are included in the breakage function model. In this way, the model accounts for the material dependency of the breakage functions. Also, the breakage functions are dependent on the size of the feed material. The results of the impact comminution experiments were found to be the same for impact angles of both $60°$ and $90°$, as long as only the velocity component normal to the impact plane is considered. Still, it cannot be excluded that the horizontal velocity component becomes important at smaller impact angles.

Nanoindentation experiments were done in order to characterise the mechanical properties of the substances. Hardness, Young's modulus and fracture toughness were determined by this method. These properties were compared to the breakage parameters f^*_{Mat} and $x \cdot W^*_{m,min}$ derived from single particle comminution. It could be shown that a correlation exists between the breakage parameters and the brittleness index H/K_c. This proves that both f^*_{Mat} and $x \cdot W^*_{m,min}$ are influenced by the plastic properties of the materials, and shows a way to predict these quantities from measurements of the mechanical properties. Two equations are given to derive the breakage parameters from mechanical properties. It could also be shown that nanoindentation is a powerful tool to determine the mechanical properties of particulate materials. Results from nanoindentation can be used to estimate the milling performance of the materials not only qualitatively, but even quantitatively. This means also a remarkable reduction of the amount of substance needed: while the single particle impact tests still need about 50 g of substance, indentations can be performed with only a few crystals.

In bulk milling experiments, Compound B was identified to exhibit an extremely sticky behaviour, it easily sticks to the walls of the equipment and clogs the mill after a short time. In experiments characterising adhesive interactions and powder flowability, this substance was used as reference to detect critical influences concerning the adhesive behaviour of a material. In shear tests, the flowability was found to be only a little lower than that of the other studied substances. The friction on a smooth steel wall was found to be significantly higher for Compound B than for the other substances, while on PTFE, all sample materials show a similar wall friction.

Atomic force microscopy was used to characterise the adhesion of different particulate organic substances, both on themselves and on different wall samples, and at different levels of relative humidity. The results were found to vary only slightly for different substances. Van-der-Waals forces are similar, also the adhesion at different levels of relative humidity is changing basically to the same extent for the various substances. No clue was found to identify an exceptionally strong adhesive behaviour of Compound B. The results obtained from AFM, namely the relative adhesion forces on steel and on PTFE, can not be transferred to the shear experiments, because different types and different levels of stress occur. During milling, even higher stresses can be expected. Since no intrinsic stickiness of Compound B was found at low load levels, it seems likely that the stickiness is mainly caused by deformations that happen at higher loads. The surface of the crystals is very smooth already from the beginning; this smoothness and also elastic-plastic deformations occurring under stress enhance a good contact to other adhesion partners and eventually result in clogging.

AFM measurements between different organic substances and on different crystal faces of lactose show that adhesion depends strongly on the local surface properties of both adhesion partners and their interaction. This is of importance especially when polar surface groups are involved on both contacting surfaces. Another aspect that has not been assessed in this work is the electrostatic behaviour of the materials. Electrostatic charging was observed in some cases, but the interactions induced by charging were not studied in detail.

With inverse gas chromatography, the dispersive component of the surface energy was measured. Differences of the surface energy of powders before and after milling were observed, which is probably due to a partly amorphisation of the particle surfaces during milling. The largest difference was found for the substance with the poorest comminution behaviour. However, a broader data basis consisting of more substances is required in order to make proper conclusions about whether the change of surface properties upon milling, as measured by iGC, can serve as a measure for the grinding performance of a substance.

Besides these findings, some questions remain: The presented model for the breakage functions was derived from experiments where breakage probabilities below ~ 90 % were obtained. There are some indications that breakage functions may be different when higher extents of breakage are achieved. Thus, further experiments will be necessary at very high impact velocities, in order to get a detailed picture how the breakage functions will behave when practically complete breakage is achieved.

The correlation between brittleness index and the breakage parameters of a material could be shown, but at present it is not clear how these properties are correlated in detail to the structure of the material itself. In fact, is is not even clear yet whether the correlation presented here is only applicable to crystalline structures as they were studied here, or whether it is generally applicable also to amorphous materials or structured compound materials. To address this question, more studies need to be done with other materials, and a close look needs to be taken at the inner structure of these materials. Another very important question is, whether the results shown here, which were derived from impact comminution experiments, are also applicable to other kinds of stressing, such as crushing or comminution in a stirred media mill, where compressive stressing and shear stressing take place.

Since it was found that bulk adhesive behaviour is not properly accessible by AFM, the question remains how the problem of scaling can be addressed adequately. Obviously, experiments need to be done at similar stressing conditions as in a mill - e.g. shear tests at higher normal stresses. Also, close examination of layers of scaled material should reveal details about the structure of these layers and thus give insight about the scaling process of these materials.

As another practical outcome of this work, some advice may be given as how to improve the milling behaviour of a substance: in the first case, i. e. for a substance with unfavourable mechanical properties, the breakage behaviour may be enhanced already during particle generation: e. g. by choosing appropriate crystallisation conditions, another crystal structure may be obtained, or defects may be induced in the crystal lattice that enhance the brittleness of the particles. For the second case, i. e. poor transport properties of the material: this issue has not been totally clarified yet, but there are hints that these properties can be improved also by the crystallisation conditions: inducing rough or uneven surfaces helps reducing adhesion forces, also the formation of irregular aggregate structures (that will be easier to comminute) can improve the transport properties.

6. NOMENCLATURE

A	[m²]	contact area
A	[J]	Hamaker constant
a	[m]	indent diagonal
a	[m²]	surface area per adsorbate molecule
a_1	[-]	mixing ratio of the bimodal log-normal distribution
B(x,y)	[-]	cumulative breakage function
b_{ij}	[-]	breakage coefficients
c	[m]	crack radius (c = l + a)
c, c_1, c_2	[-]	proportionality constants
c_{min}	[m]	critical flaw size
D	[m]	characteristic length
D	[-]	relative density
D_0	[-]	initial relative density
d_f	[m]	fringe spacing
d_R	[m]	rotor diameter
E	[Pa]	Young's modulus
E_{fract}	[J/m³]	fracture energy
E_r	[Pa]	reduced modulus
E_v	[J/m³]	volume-specific energy
F	[N]	force
f	[1/s]	frequency
ff_c	[-]	flowability
f_{Mat}	[kg/(J·m)]	material parameter

f^*_{Mat}	[kg/(J·m)]	adjusted material parameter
f_{Red}	[-]	velocity reduction factor
F_{rel}	[-]	relative adhesion force
F_{self}	[N]	adhesion force on substrate of the same material
F_{wall}	[N]	adhesion force on wall
G	[Pa]	shear modulus
ΔG^0	[J]	free enthalpy of adsorption
g_1, g_2		functions
H	[Pa]	hardness
H	[mol/(kg·Pa)]	Henry coefficient
$H_{S,BET}$	[mol/(m²·Pa)]	surface related Henry coefficient
h	[m]	indentation depth
h_p	[m]	plastic indentation depth
j	[-]	James-Martin correction factor
K	[1/Pa]	Heckel constant
K_c	[Pa·m$^{-1/2}$]	fracture toughness
K_{et}	[Pa]	elasticity parameter
K_{IC}	[Pa·m$^{-1/2}$]	stress intensity factor
k	[-]	number of impacts
$k_i(n)$	[-]	normalised stress intensity factor after Ouchterlony
l	[m]	crack length
l_i	[m]	initial flaw size
m	[kg]	mass
m	[-]	parameter of the GGS distribution
m_i	[kg]	mass in interval i
m_s	[kg]	solid mass
N	[-]	Avogadro number
n	[-]	parameter of the RRSB distribution
n	[1/s]	number of revolutions

n_{ads}	[mol/kg]	adsorbate loading
P	[N]	indentation load
P_y	[Pa]	yield pressure
p	[Pa]	pressure
p_{in}	[Pa]	pressure at column inlet
p_{out}	[Pa]	pressure at column outlet
$Q_3(x)$	[-]	cumulative particle size distribution
$q_3(x)$	[1/m]	density particle size distribution
$q_B(x)$	[1/m]	breakage function (density distribution)
$q_F(x)$	[1/m]	density distribution of feed material
$q_P(x)$	[1/m]	density distribution of comminuted sample
R	[m]	radius
R	[J/m²]	crack resistance
R	[J/(mol·K)]	universal gas constant
R^2	[-]	empirical correlation coefficient
r	[m]	asperity radius
rms	[m]	root mean square roughness
S	[-]	breakage probability
S_{app}	[-]	apparent breakage probability
S_{BET}	[m²/kg]	BET surface
S_i	[-]	breakage probability of size interval i
S_{nozzle}	[-]	breakage probability in the dispersion nozzle
S_{plate}	[-]	breakage probability on the impact plate
S_v	[m^{-1}]	specific surface area
T	[K]	temperature
t_0	[s]	dead time
t_R	[s]	retention time
V	[m³]	volume
$\dot{V}$	[m³/s]	volumetric flow rate

141

V_a	[m³]	attrited volume
V_N	[m³]	net retention volume
v	[m/s]	velocity
v_d	[m/s]	velocity of deformation
v_{el}	[m/s]	elastic wave propagation velocity
v_{fract}	[m/s]	velocity of crack propagation
v_{gas}	[m/s]	gas velocity
v_P	[m/s]	particle velocity
W_A	[J]	energy of adhesion
W_c	[J]	work of compression
$W^*_{m,eff}$	[J/kg]	mass-specific effective impact energy
$W_{m,kin}$	[J/kg]	mass-specific kinetic energy
$W^*_{m,kin}$	[J/kg]	adjusted mass-specific kinetic energy that is available for breakage
$W_{m,min}$	[J/kg]	mass-specific threshold energy for particle breakage
$W^*_{m,min}$	[J/kg]	adjusted mass-specific threshold energy for particle breakage
$W_{m,max}$	[J/kg]	maximum energy available for comminution ($=W^*_{m,kin}$)
x	[m]	particle size
Δx	[m]	interval size
x^*	[m]	parameter of the attenuated GGS distribution
x'	[m]	parameter of the RRSB distribution
x_0	[m]	initial particle size (mean diameter)
x_1, x_2	[m]	deformations
x_{50}	[m]	median diameter of the log-normal distribution
x_A, x_B	[m]	median diameters of bimodal log-normal distributions
x_b	[-]	calibration constant for Berkovich indenter
x_v	[-]	calibration constant for Vickers indenter
x_{crit}	[m]	critical size for brittle-ductile transition
x_{max}	[m]	parameter of the GGS distribution

Y	[Pa]	yield strength
y	[m]	initial particle size
Δz	[m]	jump-out distance
z_0	[m]	minimum contact distance

Greek letters

α	[-]	test factor
α, α^*	[-]	proportionality constants
α_1, α_2	[$N^{-2/3}m$]	elasticity parameter
β	[-]	geometric correction factor for Berkovich indenter
β_{max}	[J/m²]	maximum fracture energy
γ	[J/m²]	surface energy
γ_L^d	[J/m²]	dispersive component of surface energy of the solid
γ_S^d	[J/m²]	dispersive component of surface energy of the liquid
δ	[$Pa^{1/2}$]	solubility parameter
ε	[-]	porosity
ε_0	[-]	initial porosity
η	[Pa·s]	dynamic viscosity
θ	[-]	semiapical angle of indenter
λ	[m]	wavelength
λ	[-]	Maugis parameter
ν	[-]	Poisson ratio
ξ	[-]	fractional mass loss
Π_i	[-]	pi products (dimensional analysis)
ρ	[kg/m³]	density
ρ_b	[kg/m³]	bulk density
ρ_p	[kg/m³]	particle density
σ	[-]	standard deviation
σ	[Pa]	normal stress

σ_0	[-]	standard deviation of initial size distribution
σ_0	[Pa]	yield strength
σ_1	[Pa]	major consolidation stress
σ_A, σ_B	[-]	standard deviations of the bimodal log-normal distributions
σ_B	[Pa]	breaking stress
σ_c	[Pa]	unconfined yield strength
σ_w	[Pa]	normal stress in wall shear experiments
τ	[Pa]	shear stress
τ_w	[Pa]	wall shear stress
φ	[-]	impact angle
φ	[-]	wall friction angle
φ	[-]	half-angle of intersection
Ψ	[-]	Stokes parameter
Ψ^*	[-]	modified Stokes parameter

7. REFERENCES

Austin L. G.: *Introduction to the Mathematical Description of Grinding as a Rate Process*, Powder Technology, 5 **(1971)** 1-17

Baie S. B., Newton J. M., Podczeck F.: *The Characterization of the Mechanical Properties of Pharmaceutical Materials*, European Journal of Pharmaceutics and Biopharmaceutics, 43 **(1996)** 2, 138-1413

Becker M., Kwade A., Schwedes J.: *Stress Intensity in Stirred Media Mills and its Effect on Specific Energy Requirement*, International Journal of Mineral Processing, 61 **(2001)** 189-208

Begat P., Morton D. A. V., Staniforth J. N., Price R.: *The Cohesive-Adhesive Balances in Dry Powder Inhaler Formulations I: Direct Quantification by Atomic Force Microscopy*, Pharmaceutical Research, 21 **(2004)** 9, 1591-1597

Bernotat S., Schönert K.: *Size Reduction*, in: Ullmann's Encyclopedia of Industrial Chemistry, 7th ed., Wiley VCH **(2000)**

Blendell J. E.: PhD thesis, Massachussetts Institute of Technology **(1979)**

Bond F.: *The Third Theory of Comminution*, Transactions of the Society of Mining Engineers of AIME, 193 **(1952)** 484

Broadbent S. R., Callcott T. G.: *A Matrix Analysis of Processes involving Particle Assemblies*, Philosophical Transactions of the Royal Society of London, Series A, 249 **(1956)** 960, 99-123

Buss B.: *Verwendung mehrparametriger logarithmischer Normalverteilungen zur Darstellung der Korngrößenverteilungen von Zerkleinerungsprodukten*, Freiberger Forschungshefte A, 560 **(1976)** 7-28

Chan H. K., Doelker E.: *Polymorphic Transformation of some Drugs under Compression*, Drug Development and Industrial Pharmacy, 11 (**1985**) 2-3, 315-332

Courtney T. H.: *Mechanical Behavior of Materials, 2ⁿᵈ ed.*, McGraw-Hill (**2000**)

Crabtree D. D., Kinasevich R. S., Mular A. L., Meloy T. P., Fuerstenau D. W.: *Mechanisms of Size Reduction in Comminution Systems. Part I. Impact, Abrasion, and Chipping Grinding*, Transactions of the American Institute of Mining, Metallurgical & Petroleum Engineers, 229 (**1964**) 201-206

Derjaguin B. V., Muller V. M., Toporov Y. R.: *Effect of Contact Deformations on the Adhesion of Particles*, Journal of Colloid and Interface Science, 53 (**1975**) 2, 314-326

Doerner M. F., Nix W. D.: *A Method for Interpreting the Data from Depth-Sensing Indentation Instruments*, Journal of Materials Research, 1 (**1986**) 4, 601-609

Ducker W. A., Senden T. J., Pashley R. M.: *Direct Measurement of Colloidal Forces using an Atomic Force Microscope*, Nature, 353 (**1991**) 239-241

Dukino R. D., Swain M. V.: *Comparative Measurement of Indentation Fracture Toughness with Berkovich and Vickers Indenters*, Journal of the American Ceramic Society, 75 (**1992**) 12, 3299-3304

Duncan-Hewitt W. C., Weatherly G. C.: *Evaluating the Hardness, Young's modulus and Fracture Toughness of some Pharmaceutical Crystals using Microindentation Techniques*, Journal of Materials Science Letters, 8 (**1989**) 1350-1352

Durst F., Melling A., Whitlaw J. H.: *Principles and Practice of Laser-Doppler Anemometry, 2ⁿᵈ ed.*, Academic Press, London (**1981**)

Eliasson B., Dändliker R.: *A Theoretical Analysis of Laser Doppler Flowmeters*, Optica Acta, 21 (**1974**) 2, 119-149

Epstein B.: *Logarithmico-Normal Distribution in Breakage of Solids*, Industrial & Engineering Chemistry, 40 (**1948**) 12, 2289-2291

Evans A. G., Charles A.: *Fracture Toughness Determinations by Indentation*, Journal of the American Ceramic Society, 59 (**1976**) 371-372

Evans A. G.: *Fracture Toughness: The Role of Indentation Techniques*, in: Fracture Mechanics Applied to Brittle Materials, ASTM STP678, S. W. Freiman (ed.), ASTM **(1979)** 112-135

Fagan P. G., Harding V. D., Norman G. T., Arteaga P. A., Ghadiri M.: *Comparative Study of the Mechanical Behavior of Alpha-Lactose from Measurements on Compacted Beam Specimens, Controlled Powder Compression, and Single Crystals*, Proc. 5th World Congress on Chem. Eng. / 2nd Particle Technol. Forum, San Diego **(1996)** 590-595

Fischer-Cripps A. C.: *Nanoindentation, 2nd ed.*, Springer, New York **(2004)**

Fowkes F. M.: *Attractive Forces at Interfaces*, Industrial & Engineering Chemistry, 56 **(1964)** 40-52

Frye L., Peukert W.: *Attrition of Bulk Solids in Pneumatic Conveying: Mechanisms and Material Properties*, Particle Science & Technology, 20 **(2002)** 4, 267-282

Gabaude C. M. D., Guillot M., Gautier J.-C., Saudemon P., Chulia D.: *Effects of True Density, Compacted Mass, Compression Speed, and Punch Deformation on the Mean Yield Pressure*, Journal of Pharmaceutical Sciences, 88 **(1999)** 7, 725-730

Gahn C.: *Die Festigkeit von Kristallen und ihr Einfluss auf die Kinetik in Suspensionskristallisatoren*, PhD thesis, Technical University Munich **(1997)**

Gebelein H.: *Beiträge zum Problem der Kornverteilungen*, Chemie Ingenieur Technik, 28 **(1956)** 12, 773-782

Ghadiri M., Zhang Z.: *Impact Attrition of Particulate Solids: Part 1: A Theoretical Model of Chipping*, Chemical Engineering Science, 57 **(2002)** 3659-3669

Gommeren H. J. C., Heitzmann D. A., Moolenaar J. A. C., Scarlett B.: *Modelling and Control of a Jet Mill Plant*, Powder Technology, 108 **(2000)** 147-154

Götzinger M., Peukert W.: *Particle Adhesion Force Distributions on Rough Surfaces*, Langmuir, 20 **(2004)** 5298-5303

Götzinger M.: *Zur Charakterisierung von Wechselwirkungen partikulärer Oberflächen*, PhD thesis, Universität Erlangen-Nürnberg (**2005**)

Hagan J. T.: *Micromechanics of Crack Nucleation during Indentations*, Journal of Materials Science, 14 (**1979**) 2975-2980

Hamaker H. C.: *The London-van der Waals Attraction between Spherical Particles*, Physica, 4 (**1937**) 1058-1072

Hanisch J., Schubert H.: *Comminution of Irregularly Shaped Particles by Slow Compression: Interpretation of the Size Distributions of Progeny Particles as Mixed Distributions*, Particle Characterization, 1 (**1984**) 74-77

Heckel R. W.: *Density-Pressure Relationships in Powder Compaction*, Transactions of the American Institute of Mining, Metallurgical & Petroleum Engineers, 221 (**1961a**) 671-675

Heckel R. W.: *An Analysis of Powder Compaction Phenomena*, Transactions of the American Institute of Mining, Metallurgical & Petroleum Engineers, 221 (**1961b**) 1001-1008

Hencky H.: *Über einige statisch bestimmte Fälle des Gleichgewichts in plastischen Körpern*, Zeitschrift für Angewandte Mathematik und Mechanik, 3 (**1923**) 241-251

Hertz H.: *Über die Berührung fester elastischer Körper*, Journal für die reine und angewandte Mathematik, 92 (**1881**) 156-171

Israelachvili J.: *Intermolecular and Surface Forces*, Academic Press, San Diego (**1992**)

Jenike A. W.: *Bull. No. 123*, University of Utah, Salt Lake City (**1964**)

Johnson K. L., Kendall K., Roberts A. D.: *Surface Energy and the Contact of Elastic Solids*, Proceedings of the Royal Society of London, Series A, 324 (**1971**) 301-313

Jones R., Pollock H. M., Cleaver J. A. S., Hodges C. S.: *Adhesion Forces between Glass and Silicon Surfaces in Air Studied by AFM: Effects of Relative Humidity, Particle Size, Roughness and Surface Treatment*, Langmuir, 18 (**2002**) 8045-8055

Kappl M., Butt H.-J.: *The Colloidal Probe Technique and its Application to Adhesion Force Measurements*, Particle & Particle Systems Characterization, 19 **(2002)** 129-143

Kendall K.: *The Impossibility of Comminuting Small Particles by Compression*, Nature, 272 **(1978)** 710-711

Kick F.: *Das Gesetz der proportionalen Widerstände und seine Anwendung*, A. Felix, Leipzig **(1885)**

Klotz K.: *On the Fragment Size Distribution of a Very Thin Infinite Brittle Plate*, Crystal Research and Technology, 16 **(1981)** 3, K25-K27

Klotz K., Schubert H.: *Crushing of Single Irregularly Shaped Particles by Compression: Size Distribution of Progeny Particles*, Powder Technology, 32 **(1982)** 129-137

Krupp H.: *Particle Adhesion - Theory and Experiment*, Advances in Colloid and Interface Science, 1 **(1967)** 111-239

Kwan C. C., Chen Y. Q., Ding Y. L., Papadopoulos D. G., Bentham A. C., Ghadiri M.: *Development of a Novel Approach towards Predicting the Milling Behaviour of Pharmaceutical Powders*, European Journal of Pharmaceutical Sciences, 23 **(2004)** 327-336

Larsson I., Kristensen H. G.: *Comminution of a Brittle/Ductile Material in a Micros Ring Mill*, Powder Technology, 107 **(2000)** 175-178

Lawn B. R., Evans A. G.: *A Model for Crack Initiation in Elastic/Plastic Indentation Fields*, Journal of Materials Science, 12 **(1977)** 2195-2199

Lawn B. R., Marshall D. B.: *Hardness, Toughness, and Brittleness: An Indentation Analysis*, Journal of the American Ceramic Society, 62 **(1979)** 07. Aug, 347-350

Lecoq O., Chouteau N., Mebtoul M., Large J.-F., Guigon P.: *Fragmentation by High Velocity Impact on a Target: a Material Grindability Test*, Powder Technology, 133 **(2003)** 1-3, 113-124

Löffler F., Raasch J.: *Grundlagen der mechanischen Verfahrenstechnik*, Vieweg, Braunschweig **(1992)**

Louey M.D., Mulvaney P., Stewart P. J.: *Characterisation of Adhesional Properties of Lactose Carriers using Atomic Force Microscopy*, Journal of Pharmaceutical and Biomedical Analysis, 25 **(2001)** 559-567

Maugis D.: *Adhesion of Spheres: The JKR-DMT Transition using a Dugdale Model*. Jounal of Colloid and Interface Science, 150 **(1992)** 1, 243-269

Maurer S.: *Prediction of Single-Component Adsorption Equilibria*, PhD thesis, Technical University Munich **(2000)**

McCook N. L., Burris D. L., Bourne G. R., Steffens J., Hanrahan J. R., Sawyer W. G.: *Wear Resistant Solid Lubricant Coating made from PTFE and Epoxy*, Tribology Letters, 18 **(2005)** 1, 119-124

Meier M., John E., Wieckhusen D., Wirth W., Peukert W.: *Characterization of the grinding behaviour in a single particle impact device: Studies on pharmaceutical powders*, European Journal of Pharmaceutical Sciences 34 **(2008)** 45-55

Meier M., John E., Wieckhusen D., Wirth W., Peukert W.: *Generally applicable breakage functions derived from single particle comminution data*, Powder Technology 194 **(2009a)** 33-41

Meier M., John E., Wieckhusen D., Wirth W., Peukert W.: *Influence of mechanical properties on impact fracture: Prediction of the milling behaviour of pharmaceutical powders by nanoindentation*, Powder Technology 188 **(2009b)** 301-313

Menzel U.: *Theoretische und experimentelle Untersuchungen an einer Prallplatten-Strahlmühle*, PhD Thesis, Technische Universität Clausthal **(1987)**

Midoux N., Hošek P., Pailleres L., Authelin J. R.: *Micronization of Pharmaceutical Substances in a Spiral Jet Mill*, Powder Technology, 104 **(1999)** 113-120

Müller F., Polke R., Schäfer M.: *Model-based Evaluation of Grinding Experiments*, Powder Technology, 105 **(1999)** 243-249

Newell H. E., Buckton G., Butler D. A., Thielmann F., Williams D. R.: *The Use of Inverse Phase Gas Chromatography to Measure the Surface Energy of Crystalline, Amorphous, and Recently Milled Lactose*, Pharmaceutical Research, 18 **(2001)** 5, 662-666

Nix W. D., Gao H.: *Indentation Size Effects in Crystalline Materials: a Law for Strain Gradient Plasticity*, Journal of the Mechanics & Physics of Solids, 46 **(1998)** 3, 411-425

Oliver W. C., Pharr G. M.: *An Improved Technique for Determining Hardness and Elastic Modulus using Load and Displacemnt Sensing Indentation Experiments*, Journal of Materials Research, 7 **(1992)** 6, 1564-1583

Ouchterlony F.: *Stress Intensity Factors for the Expansion Loaded Star Crack*, Engineering Fracture Mechanics, 8 **(1976)** 447-448

Palmqvist S.: *Metod att Bestämma Segheten hos Spröda Material, särskilt Hardmetaller*, Jernkontorets Annaler, 141 **(1957)** 5, 300-307

Palmqvist S.: *Rißbildungsarbeit bei Vickers-Eindrücken als Maß für die Zähigkeit von Hartmetallen*, Archiv für das Eisenhüttenwesen, 33 **(1962)** 9, 629-634

Paronen P., Juslin M.: *Compressional Characteristics of Four Starches*, Journal of Pharmacy and Pharmacology, 35 **(1983)** 627-635

Paronen P.: *Heckel Plots as Indicators of Elastic Properties of Pharmaceuticals*, Drug Development & Industrial Pharmacy, 12 **(1986)** Nov 13, 1903-1912

Pedersen S., Kristensen H. G.: *Change in Crystal Density of Acetylsalicylic Acid during Compaction*, S.T.P. Pharma Sciences, 4 **(1994)** 3, 201-206

Planinsek O., Zadnik J., Rozman S., Kunaver M., Dreu R., Srcic S.: *Influence of Inverse Gas Chromatography Measurement Conditions on Surface Energy Parameters of Lactose Monohydrate*, International Journal of Pharmaceutics, 256 **(2003)** 17-23

Podczeck F., Newton J. M., James M. B.: *Assessment of Adhesion and Autoadhesion Forces between Particles and Surfaces: I. The Investigation of Autoadhesion Phenomena of Salmeterol Xinafoate and Lactose Monohydrate Particles using Compacted Powder Surfaces*, Journal of Adhesion Science & Technology, 8, 12 **(1994)** 1459-1472

Podczeck F., Newton J. M., James M. B.: *Adhesion and Autoadhesion Measurements of Micronized Particles of Pharmaceutical Powders to Compacted Powder Surfaces*, Chemical & Pharmaceutical Bulletin, 43, 11 **(1995)** 1953-1957

Podczeck F., Newton J. M., James M. B.: *The Adhesion Force of Micronized Salmeterol Xinafoate Particles to Pharmaceutically Relevant Surface Materials*, Journal of Physics D: Applied Physics, 29 **(1996)** 1878-1884

Podczeck F.: *Investigations into the Fracture Mechanics of Acetylsalicylic Acid and Lactose Monohydrate*, Journal of Materials Science, 36 **(2001)** 4687-4693

Ponton C. B., Rawlings R. D.: *Vickers Indentation Fracture Toughness Test Part 1: Review of Literature and Formulation of Standardised Indentation Toughness Equations*, Materials Science & Technology, 5 **(1989a)** 9, 865-872

Ponton C. B., Rawlings R. D.: *Vickers Indentation Fracture Toughness Test Part 2: Application and Critical Evaluation of Standardised Indentation Toughness Equations*, Materials Science & Technology, 5 **(1989b)** 10, 961-976

Prasad K. V. R., Sheen D. B., Sherwood J. N.: *Fracture Property Studies of Paracetamol Single Crystals using Microindentation Techniques*, Pharmaceutical Research, 18 **(2001)** 6, 867-872

Price R., Young P. M., Edge S., Staniforth J. N.: *The Influence of Relative Humidity on Particulate Interactions in Carrier-Based Dry Powder Inhaler Formulations*, International Journal of Pharmaceutics, 246 **(2002)** 47-59

Puttick K. E.: *The Correlation of Fracture Transitions*, Journal of Physics D: Applied Physics, 13 **(1980)** 2249-2262

Rabinovich Y. I., Adler J. J., Ata A., Singh R. K., Moudgil B. M.: *Adhesion between Nanoscale Rough Surfaces – I. Role of Asperity Geometry*, Journal of Colloid and Interface Science, 232 **(2000)** 10-16

von Rittinger P.: *Lehrbuch der Aufbereitungskunde*, Ernst & Korn, Berlin **(1867)**

Roberts R. J., Rowe R. C., York P.: *The Measurement of the Critical Stress Intensity Factor (K_{IC}) of Pharmaceutical Powders using Three-Point Single Edge Notched Beam (SENB) Testing*, International Journal of Pharmaceutics, 91 **(1993)** 173-182

Roberts R. J., Rowe R. C., York P.: *The Relationship between the Fracture Properties, Tensile Strength and Critical Stress Intensity Factor of Organic Solids and their Molecular Structure*, International Journal of Pharmaceutics, 125 **(1995)** 157-162

Ruck B.: *Laser-Doppler-Anemometrie*, AT-Fachverlag, Stuttgart **(1987)**

Rumpf H.: *Physical Aspects of Comminution and New Formulation of a Law of Comminution*, Powder Technology, 7 **(1973)** 145-159

Rumpf H.: *Mechanische Verfahrenstechnik*, Hanser Verlag, München **(1975)**

Salman A. D., Gorham D. A., Verba A.: *A Study of Solid Particle Failure under Normal and Oblique Impact*, Wear, 186-187 **(1995)** 92-98

Schlichting H.: *Boundary Layer Theory, 7th ed.*, McGraw-Hill, New York **(1987)**

Schönert K., Marktscheffel M.: *Liberation of Composite Particles by Single Particle Compression, Shear and Impact Loading*, 6. European Symposium on Comminution, Nürnberg **(1986)** 29-45

Schultz J., Lavielle L., Martin C.: *The Role of the Interface in Carbon Fibre-Epoxy Composites*, Journal of Adhesion, 23 **(1987)** 45-60

Schulze D.: *Entwicklung und Anwendung eines neuartigen Ringschergerätes / Development and Application of a Novel Ring Shear Tester*, Aufbereitungs-Technik/Mineral Processing 35 **(1994)** 10, 524-535

Schwedes J., Schulze D.: *Measurement of Flow Properties of Bulk Solids*, Powder Technology, 61 **(1990)** 59-68

Sindel U., Zimmermann I.: *Measurement of Interaction Forces between Individual Powder Particles using an Atomic Force Microscope*, Powder Technology, 117 **(2001)** 247-254

Sonnergaard J. M.: *A Critical Evaluation of the Heckel Equation*, International Journal of Pharmaceutics, 193 **(1999)** 63-71

Stephan K., Mayinger F.: *Thermodynamik: Grundlagen und technische Anwendungen, 15. Auflage*, Springer-Verlag Berlin/Heidelberg **(1998)**

Sun C., Grant D. J.: *Influence of Elastic Deformation of Particles on Heckel Analysis*, Pharmaceutical Development & Technology, 6 **(2001)** 2, 193-200

Tabor D.: *Indentation Hardness and its Measurement: some Cautionary Comments*, in: Microindentation Techniques in Materials Science and Engineering, ASTM STP 889, P. J. Blau, B. R. Lawn (ed.), ASTM **(1986)** 129-159

Taylor L. J., Papadopoulos D. G., Dunn P. J., Bentham A. C., Mitchell J. C., Snowden M. J.: *Mechanical Characterisation of Powders using Nanoindentation*, Powder Technology, 143-144 **(2004a)** 179-185

Taylor L. J., Papadopoulos D. G., Dunn P. J., Bentham A. C., Dawson N. J., Mitchell J. C., Snowden M. J.: *Predicitive Milling of Pharmaceutical Materials Using Nanoindentation of Single Crystals*, Organic Process Research & Development, 8 **(2004b)** 674-679

Ticehurst M. D., York P., Rowe R. C., Dwivedi S. K.: *Characterisation of the Surface Properties of α-Lactose Monohydrate with Inverse Gas Chromatography, used to detect Batch Variation*, International Journal of Pharmaceutics, 141 **(1996)** 93-99

Tsukada M., Irie R., Yonemuchi Y., Noda R., Kamiya H., Watanabe W., Kauppinen E. I.: *Adhesion Force Measurement of a DPI Size Pharmaceutical Particle by Colloid Probe Atomic Force Microscopy*, Powder Technology, 141 **(2004)** 262-269

de Vegt O., Vromans H., Faassen F., van der Voort Maarschalk K.: *Milling of Organic Solids in a Jet Mill. Part 1: Determination of the Selection Function and Related Mechanical Material Properties*, Particle & Particle Systems Characterization, 22 **(2005a)** 133-140

de Vegt O., Vromans H., Faassen F., van der Voort Maarschalk K.: *Milling of Organic Solids in a Jet Mill. Part 2: Checking the Validity of the Predicted Rate of Breakage Function*, Particle & Particle Systems Characterization, 22 **(2005b)** 261-267

Vogel L., Peukert W.: *Separation of the Influences of Material and Machine in Impact Comminution - Modelling with Population Balances*, Minerals Processing, 43 **(2002)** 8, 19-30

Vogel L., Peukert W.: *Breakage Behaviour of Different Materials – Construction of a Mastercurve for the Breakage Probability*, Powder Technology, 129 **(2003)** 101-110

Vogel L., Peukert W.: *From Single Particle Impact Behaviour to Modelling of Impact Mills*, Chemical Engineering Science, 60 **(2005)** 5164-5176

Willing G. A., Ibrahim T. H., Etzler F. M., Neuman R. D.: *New Approach to the Study of Particle-Surface Adhesion Using Atomic Force Microscopy*, Journal of Colloidal and Interface Science, 226 **(2000)** 185-188

Willing G. A., Burk T. R., Etzler F. M., Neuman R. D.: *Adhesion of Pharmaceutical Particles to Gelatin Capsules having Variable Surface Physicochemical Properties: Evaluation using a Combination of Scanning Probe Microscopy Techniques*, Colloids and Surfaces A: Physicochemical and Engineering Aspects, 193 **(2001)** 117-127

Young P. M., Price R., Tobyn M. J., Buttrum M., Dey F.: *Investigation into the Effect of Humidity on Drug-Drug Interacions Using the Atomic Force Microscope*, Journal of Pharmaceutical Sciences, 92, 4 **(2003)** 815-822

Yuregir K. R., Ghadiri M., Clift R.: *Observations on Impact Attrition of Granular Solids*, Powder Technology, 49 **(1986)** 53-57

Zügner S.: *Untersuchungen zum elastisch-plastischen Verhalten von Kristalloberflächen mittels Kraft-Eindringtiefen-Verfahren*, PhD thesis, Universität Würzburg **(2002)**

Zügner S., Marquardt K., Zimmermann I.: *Influence of Nanomechanical Crystal Properties on the Comminution Process of Particulate Solids in Spiral Jet Mills*, European Journal of Pharmaceutics and Biopharmaceutics, 62 **(2006)** 194-201

APPENDIX

A. Fit qualities of breakage probabilities

Table A-1 Fit qualities of breakage probabilities obtained from laser diffraction data

Substance	$v_{particle}$ [m/s]	S_{app} (LD) [-]	R^2 [-]
Acetylsalicylic acid	26.4	0.324	0.967
	31.1	0.413	0.94
	35.2	0.530	0.941
	39.4	0.529	0.935
	43.4	0.820	0.939
α-lactose	36.0	0.136	0.959
monohydrate	56.0	0.333	0.941
	64.3	0.407	0.931
	78.3	0.636	0.944
	90.6	0.686	0.939
Citric acid	23.8	0.343	0.959
monohydrate	28.4	0.519	0.95
	30.2	0.694	0.936
	34.3	0.887	0.966
Compound A	96.2	0.279	0.93
	97.5	0.287	0.9
Sucrose	25.7	0.141	0.976
	31.7	0.342	0.975
	36.1	0.473	0.977
	39.1	0.556	0.923
	43.8	0.582	0.875
	57.4	0.808	0.953
	62.6	0.829	0.946

Table A-1 (continued)

Substance	$v_{particle}$ [m/s]	S_{app} (LD) [-]	R^2 [-]
Ascorbic acid	21.8	0.301	0.951
	25.3	0.484	0.916
	32.1	0.572	0.871
	37.5	0.678	0.887
	44.3	0.796	0.931
	50.9	0.889	0.965
Glycine	20.9	0.086	0.982
	30.0	0.160	0.948
	37.2	0.271	0.882
	45.4	0.403	0.903
	52.1	0.478	0.893
	58.9	0.541	0.93
	73.4	0.730	0.934
	80.5	0.810	0.964
Tartaric acid	39.7	0.127	0.946
	48.9	0.215	0.888
	59.4	0.318	0.9
	67.5	0.432	0.987
	72.0	0.511	0.993
	75.2	0.527	0.908
	80.1	0.560	0.944
	90.6	0.626	0.853
	101.6	0.765	0.888

$v_{particle}$ = absolute velocity (impact angle not yet considered)

Impact angle was 60° for all substances, except Compound A: 75°

157

Table A-2 Fit qualities of breakage probabilities from experiments without impact plate

Substance	$v_{particle}$ [m/s]	S_{app} (LD) [-]	R^2 [-]
Acetylsalicylic acid	20.9	0.158	0.988
(old nozzle design)	25.8	0.237	0.975
	59.4	0.391	0.971
	66.4	0.462	0.957
	75.5	0.632	0.95
Acetylsalicylic acid	27.3	0.039	0.993
(new nozzle design)	33.9	0.072	0.996
	39.6	0.074	0.995
	48.2	0.129	0.985
α-lactose	52.4	0.083	0.983
monohydrate	66.3	0.083	0.994
	77.7	0.053	0.998
	89.2	0.097	0.992
Citric acid	23.8	0.068	0.999
monohydrate	30.2	0.229	0.979
	34.3	0.409	0.968
	40.9	0.442	0.976
Sucrose	25.7	0.005	0.999
	43.8	0.008	0.999
Ascorbic acid	32.1	0.048	0.92
	75.0	0.284	0.865
Glycine	20.9	0.009	0.999
	73.4	0.090	0.981
Tartaric acid	19.7	0.018	0.993
	80.1	0.040	0.977

B. Sample preparation for indentation

B.1. Glueing of particles

Glue or resin can be used to stick the particles to the supporting glass slide: large particles can be pressed into a thin glue layer, or, if the particle is large enough, glue can be applied only to the sides of the crystal to prevent it from moving sideways. With smaller particles, a particle ensemble is pressed onto a thin glue film which was stretched out using a cover slide; after drying, all particles that do not stick to the glass plate are removed using compressed air. The remaining crystals are checked under the microscope. For the glueing of larger particles, a two-compound epoxy resin (MetaFix 20, Struers GmbH, Willich, Germany) has been used in this work. For finer particles, superglue (UHU Sekundenkleber, UHU GmbH & Co. KG, Bühl, Germany) was used, because the glue film could be stretched out thinner. With other types of glue it was often found that crystals resolved in the glue, especially if the curing time of the glue was longer.

Advantages: +   The quickest method, generally applicable to all substances.

Disadvantages: -   Danger of a glue layer between particle and supporting glass slide that may affect measurements if it is too thick.

                   -   It can happen that a glue layer lies on the surface of the substrate; normally this can be detected under the microscope.

B.2. Glueing particles in their own melt

A glass slide is heated above the melting temperature of the particle. The particle is put with its flat face down onto a clean underground. The glass slide is gently pressed onto the particle from above. The particle will partly melt and stick to the glass slide after cooling down. In this work, lactose and Compound A could be prepared with this method.

Advantages: + Definitely no glue layer, and very good contact between particle and supporting glass slide.

+ Also applicable to irregularly shaped particles because a non-parallel backside of a particle can be melted away.

Disadvantages: - Only applicable to large crystals (otherwise the entire particle will melt).

- Many attempts necessary to find appropriate heating conditions for the glass slide.

B.3. Direct crystallisation

Particles can be crystallised directly on the glass slide: either, small single crystals can be grown after the method by Begat et al. (2004). Larger crystals can be grown by evaporation crystallisation: a saturated solution is allowed evaporate slowly in a beaker. Glass slides are put into the beaker, and while the solvent evaporates, crystals will form on the beaker walls and on the glass slides. Because of the slow evaporation rate, large crystals can be grown.

Advantages: + Definitely no glue layer; good contact between particle and supporting glass slide.

Disadvantages: - Orientation of the crystals is mostly the same; a parallel orientation cannot always be achieved.

- Crystals may grow into agglomerates; as a result, faces that are good for measurement may not be accessible with the indenter tip, or the faces are not properly supported (e.g. the particle is not in contact with the glass but hangs partly in the air)

General recommendation:

Try to do a direct crystallisation of large crystals. If the crystals are not directly usable for indentation, you still can choose the best crystals from this batch and fix these crystals by glueing or melting.

C. Estimation of specific energy input

For the initial milling tests described in Chapter 4.1, two different types of mill were used. As the operation principles are different, the comminution results are not directly comparable. In principle, the specific energy input would be a good basis for a comparison of different mills. However, this quantity is hard to determine in the case of a pin mill as was used in the experiments presented here.

The pin mill consists of a rotor that contains three rings of pins, and a stator alo containing three rings of pins. Comminution happens upon impacts of the particles on the pins. Since it it not known how many impacts happen to a particle on average, and under which angle the particles hit the (round) pins, it is not possible to give a proper estimate for the specific energy input.

For the spiral jet mill, the specific energy input E_{spec} can be calculated as the ratio of the energy flux that is brought into the mill by the gas flow $\dot{E}$ and the solid feed rate $\dot{M}_{powder}$ measured during the experiment:

$$E_{spec} = \frac{\dot{E}}{\dot{M}_{powder}} \tag{C.1}$$

Midoux et al. (1999) used the following approach to determine the energy flux: the nozzle flow of an ideal gas through a sonic nozzle is given by (see also Stephan and Mayinger, 1998):

$$\dot{E} = \frac{1}{2} \dot{M}_{gas} w_s^2 \tag{C.2}$$

with $\dot{M}_{gas}$ being the gas mass flow and w_s being the sonic gas velocity. These quantities can be determined from:

$$\dot{M}_{gas} = pA_{tot} \sqrt{2 \frac{M_W}{RT}} \Phi_{max} \tag{C.3}$$

where p is the applied pressure, A_{tot} is the total cross section area of the nozzles of the mill, M_w is the molecular weight, R the universal gas constant, T the temperature and Φ_{max} the effluent function:

$$\Phi_{max} = \left(\frac{2}{\kappa+1}\right)^{\frac{1}{\kappa-1}} \sqrt{\frac{\kappa}{\kappa+1}} \qquad \text{(C.4)}$$

with κ being the adiabatic coefficient. Further, the sonic gas velocity is given by

$$w_s = \sqrt{2\frac{\kappa}{\kappa+1}\frac{RT}{M_W}} \qquad \text{(C.5)}$$

and the total area of the nozzles by

$$A_{tot} = n_{nozzle}\pi\frac{d_{nozzle}^2}{4} \qquad \text{(C.6)}$$

According to this approach, the values listed in Table C-1 were obtained for the specific energy input:

Table C-1 Specific energy input of initial milling experiments

Substance	pressure	specific energy input [J/g]
Lactose S00200	2 bar	168
	4 bar	245
	6 bar	308
Lactose SV001	2 bar	105
	4 bar	210
	6 bar	315
Acetylsalicylic acid	2 bar	131
	4 bar	243
	6 bar	392
Compound A	2 bar	50
	6 bar	203
Compound B	2 bar	85
	6 bar	367

D. Control of gas velocity

For evaluation of the single particle comminution experiments, the particle velocity has to be known. This velocity can be controlled by the gas velocity. It is possible but complicated to calculate the final velocity of a particle that is accelerated along a tube from its size, density, and the flow field in the tube. For a more practical approach, the dependency of the final particle velocities on the applied air pressure was studied in detail for the new (straight) nozzle with a tube length of 600 mm. To determine the gas velocity, oil droplets were dispersed into the air stream with a nebuliser. Glass spheres of various sizes were used, the data were complemented later with data from other experiments.

Fig. D.1 shows the change of gas velocity with applied air pressure. The dependency can be well described with a power law fit equation:

$$v_{gas} /(\, m/\, s) = 81.032 \cdot (p/\, bar)^{0.5192} \tag{D.1}$$

with a correlation coefficient of $R^2 = 0.9986$. The ratio of gas velocity to particle velocity will be called the reduction factor f_{Red}:

$$f_{Red} = \frac{v_P}{v_{gas}} \tag{D.2}$$

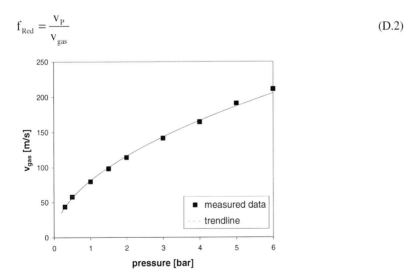

Fig. D.1 Gas velocity after the accelaration tube in dependency on the applied pressure before the tube

This ratio is related to the drag forces present during acceleration and can be described with a modified Stokes parameter Ψ^* (compare to Eq.(4.1)):

$$\Psi^* = \frac{\rho_p x^2 v_{gas}}{18\eta D} \tag{D.3}$$

The correlation between these two parameters is shown in Fig. D.2. Again, a good correlation can be given by

$$f_{Re\,d} = -0.0748\,ln(\Psi^*) + 1.2686 \tag{D.4}$$

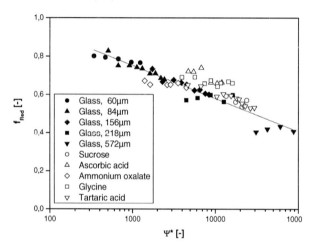

Fig. D.2 Velocity reduction factor as a function of the modified Stokes parameter Ψ^*

The data for the glass spheres of all sizes lie very near this trendline. Also for the other substances, most data are very well described by this fit. For some substances (glycine, ascorbic acid), a somewhat larger deviation exists. Most likely, this is caused by influences of the particle shape (note that the spherical glass particles exhibit less deviation than the other substances), and uncertainties in the velocity measurement. As was explained in Chapter 4.2.6, the determination of particle velocity by LDA involves a standard deviation of around 10 % in this device. Therefore, the observed deviations may be neglected, and Eqs. (D.1) to (D.4) may be applied for the determination of particle velocities as well as the directly measurement data.

E. Overview over all shear experiments

Table E-1 Results from shear tests with α-lactose monohydrate, batch S00200

sample	normal stress [Pa]	ffc [-]	bulk density [kg/m³]	σ_1 [Pa]	σ_c [Pa]
before milling	2500	3.2	710	5028	1574
pin mill 8000 rpm	2500	3.2	716	5169	1622
pin mill 13500 rpm	2500	3.2	711	5124	1590
pin mill 19000 rpm	2500	3.2	676	5214	1640
spiral jet mill, 2 bar	2500	2.0	600	5323	2704
spiral jet mill, 4 bar	2500	1.6	557	5417	3310
spiral jet mill, 6 bar	2500	1.44	515	5527	3832
before milling	5000	3.7	747	10322	2797
pin mill 8000 rpm	5000	3.8	749	10500	2766
pin mill 13500 rpm	5000	3.7	742	10538	2816
pin mill 19000 rpm	5000	3.6	710	10634	2975
spiral jet mill, 2 bar	5000	2.4	639	10556	4419
spiral jet mill, 4 bar	5000	2.0	587	10713	5411
spiral jet mill, 6 bar	5000	1.7	550	10854	6344
before milling	10000	4.3	779	20964	4878
pin mill 8000 rpm	10000	4.6	779	21035	4621
pin mill 13500 rpm	10000	4.5	772	21227	4762
pin mill 19000 rpm	10000	4.3	739	21184	4964
spiral jet mill, 2 bar	10000	2.8	674	21134	7463
spiral jet mill, 4 bar	10000	2.4	624	21457	8902
spiral jet mill, 6 bar	10000	2	583	21359	10423

Table E-2 Results from shear tests with acetylsalicylic acid

sample	normal stress [Pa]	ffc [-]	bulk density [kg/m³]	σ_1 [Pa]	σ_c [Pa]
pin mill 8000 rpm	2500	6.5	642	5073	778
pin mill 13500 rpm	2500	2.7	507	5074	1907
pin mill 19000 rpm	2500	1.8	458	5238	2851
spiral jet mill, 2 bar	2500	1.07	351	5679	5320
spiral jet mill, 4 bar	2500	1.04	284	5771	5540
spiral jet mill, 6 bar	2500	1.39	260	5998	4302
before milling	5000	n.a.	680	9927	0
pin mill 8000 rpm	5000	8.2	661	9873	1197
pin mill 13500 rpm	5000	2.9	549	9838	3450
pin mill 19000 rpm	5000	2	487	9005	4509
spiral jet mill, 2 bar	5000	1.12	379	10424	9306
spiral jet mill, 4 bar	5000	1.13	305	10746	9537
spiral jet mill, 6 bar	5000	1.7	268	11284	6593
spiral jet mill, 6 bar	5000	1.8	270	11129	6025
spiral jet mill, 6 bar	5000	1.38	280	12179	8840
pin mill 8000 rpm	10000	9.3	676	19862	2147
pin mill 13500 rpm	10000	3.6	589	18817	5159
pin mill 19000 rpm	10000	2.5	513	17088	6951
spiral jet mill, 2 bar	10000	1.24	425	19034	15305
spiral jet mill, 4 bar	10000	1.21	340	19924	16438
spiral jet mill, 6 bar	10000	1.42	311	21460	15080

Table E-3 Results from shear tests with Compound B

sample	normal stress [Pa]	ffc [-]	bulk density [kg/m³]	σ_1 [Pa]	σ_c [Pa]
before milling	2500	1.9	240	6454	3326
pin mill 19000 rpm	2500	1.6	353	6110	3879
spiral jet mill 2 bar	2500	1.61	340	6250	3873
spiral jet mill 6 bar	2500	1.58	381	6108	3865
before milling	5000	2.4	333	12328	5186
pin mill 19000 rpm	5000	2.1	392	12089	5805
spiral jet mill 2 bar	5000	2	370	12333	6022
spiral jet mill 6 bar	5000	1.8	423	11794	6571
before milling	10000	3	346	23605	7994
pin mill 19000 rpm	10000	2.6	379	22763	8770
spiral jet mill 2 bar	10000	2.6	402	24124	9341
spiral jet mill 6 bar	10000	2.2	444	22980	10338

Table E-4 Results from wall shear tests

Sample		φ on...		
		...rough steel (R_a=0.97 µm)	...smooth steel (R_a=0.21 µm)	...PTFE (R_a=1.16 µm)
Lactose S00200	before milling	28.4	21.7	23
Acetylsalicylic acid	8000 rpm	24	15.3	14.6
Compound B	before milling	30.4	21	13.3
Lactose	SJM, 4 bar	32.1	23.7	24.1
Acetylsalicylic acid	19000 rpm	29.5	19.7	15.5
Compound B	SJM, 6 bar	33.4	31.7	13.5

SJM = spiral jet mill

F. Henry coefficients from IGC

Table F-1 Henry coefficients of different adsorptives on the studied substances at 30°C

		n-hexane [10^{-9} mol/ m²/Pa]	n-heptane [10^{-9} mol/ m²/Pa]	n-octane [10^{-9} mol/ m²/Pa]	n-nonane [10^{-9} mol/ m²/Pa]	ace-tone [10^{-9} mol/ m²/Pa]	etha-nol [10^{-9} mol/ m²/Pa]
Lactose	b.m.	0.437	1.52	4.71	14.9	0.788	6.36
Lactose	SJM, 6 bar	0.439	1.30	4.10	14.3	-	-
ASA	8000 rpm	0.156	0.503	1.42	3.60	-	-
ASA	19000 rpm	0.208	0.487	1.32	3.77	0.14	0.228
ASA	SJM, 2 bar	0.207	0.594	1.71	4.86	-	-
ASA	SJM, 6 bar	0.288	0.670	1.84	5.24	-	-
Compound A, Batch 01.01<250	b.m.		0.999	3.17	9.95	-	-
Compound A, Batch 98905 MIC	SJM	-	3.19	11.2	39.5	-	0.226
Compound B	b.m.	1.94	5.58	16.9	51.4	-	-
Compound B	SJM, 2 bar	0.797	2.43	7.27	22.1	0.096	0.116
Compound B	SJM, 6 bar	0.898	2.52	7.63	22.9	-	-

b.m. = before milling
SJM = spiral jet mill

Lebenslauf

Persönliche Angaben

Name	Matthias Wilhelm Meier
Geburtsdatum	21.06.1976
Geburtsort	Waldshut-Tiengen

Schulbildung

09/83 – 07/87	Johann-Peter-Hebel-Grundschule Tiengen
09/87 – 06/96	Klettgau-Gymnasium Tiengen

Zivildienst

07/96 – 07/97	Am Krankenhaus-Spitalfond Waldshut (chirurgische Station)

Studium

10/97 – 07/03 Studium des Chemieingenieurwesens an der Universität Karlsruhe (TH); Hauptfächer: Chemische Verfahrenstechnik, Produktgestaltung; Diplomarbeitsthema: Stabilisierung von Nanopartikeln mittels „RESS"

Beruflicher Werdegang

10/03 – 07/08 Wissenschaftlicher Mitarbeiter am Lehrstuhl für Feststoff- und Grenzflächenverfahrenstechnik, Universität Erlangen-Nürnberg

Seit 08/08 Entwicklungsingenieur bei der BASF SE, Ludwigshafen, Fachbereich Partikelabscheidung und Aerosoltechnologie